The Evolution Of Man Scientifically Disproved

by

William A. Williams

The Evolution Of Man
Scientifically Disproved
by William A. Williams

ISBN: 978-93-59329-80-2

Published by

DOUBLE 9 BOOKS

2/13-B, Ansari Road
Daryaganj, New Delhi – 110002
info@double9books.com
www.double9books.com
Tel. 011-40042856

ABOUT THE AUTHOR

William A. Williams is a distinguished and accomplished author known for his contributions to the fields of history and African American studies. His extensive body of work and scholarly achievements have made him a respected figure in academia and a source of inspiration for many. Williams is particularly renowned for his insightful writings on African American history, culture, and the Civil Rights Movement. His research delves into the struggles and triumphs of African Americans throughout history, shedding light on their contributions to society and their enduring fight for civil rights and equality. As an author, Williams has penned numerous books and articles, each marked by meticulous research, critical analysis, and a deep commitment to social justice. His works often challenge prevailing narratives and offer fresh perspectives on complex historical and societal issues. Beyond his writing, William A. Williams has made significant contributions to education and scholarship. He has held academic positions at prestigious institutions and has mentored countless students and aspiring scholars, guiding them in their pursuit of knowledge and social change. Williams' dedication to the study of African American history and his advocacy for social justice continue to resonate with readers, scholars, and activists alike.

CONTENTS

Part Two
Evidence Answered

Part Three
The Soul

Introduction

Let it be understood, at the outset, that every proved theory of science is to be accepted. Only the most intense prejudice and the maddest folly would lead any one to reject the proved conclusions of science. Moreover, we should examine any new hypothesis with open minds, to see if it has in it anything truthful, helpful or advantageous. It should neither be accepted nor rejected simply because it is new. But if a theory is evidently or probably untrue, or pernicious, or at all harmful, it is to be rejected and condemned.

Some facts and objections are herein submitted to the serious seeker after truth, in the hope that a theory so out of harmony with the facts, and so destructive to the faith and the cherished hopes of man, may be completely discarded. As Evolution can not stand the acid test of mathematics, it will be repudiated by all.

We shall discuss the theory upon its merits, from a scientific standpoint, and will also demand an explanation of all facts concerned, as we have a right to do, even where they are associated with the theological and the spiritual as well as the material. We do not oppose true science but "science falsely so called." We do not ban research, but will not allow the wild vagaries of the imagination to pass as truth.

We shall not declare arbitrarily that evolution is untrue; neither will we allow scientists to decide what we shall believe. But we shall appeal to the facts, and evolution must stand or fall by the evidence. "Evolution is not to be accepted until proved." It is not yet proved and never will be.

MATHEMATICS THE ACID TEST

Every theory to which mathematics can be applied will be proved or disproved by this acid test. Figures will not lie, and mathematics will not lie even at the demand of liars. Their testimony is as clear as the mind of God. Gravitation is proved a true theory by numerous calculations, some of them the most abstruse. The Copernican theory is proved true, and the Ptolemaic theory false, by mathematical calculations. The calculations, leading to the discovery of Neptune, went far to establish the Copernican theory as well as the law of gravitation, and to disprove the Ptolemaic theory. The evolution theory, especially as applied to man, likewise is disproved by mathematics. The proof is overwhelming and decisive. Thus God makes the noble science of mathematics bear testimony in favor of the true theories and against the false theories. We shall endeavor to marshal some of the mathematical proofs against the false and pernicious theory of evolution. True theories, such as the gravitation and Copernican theories, harmonize with each other as every branch of mathematics harmonizes with every other. If evolution were true, it would harmonize with all other true theories, rather than with so many false theories.

THEORIES OF EVOLUTION

Evolution in one sense, means growth or development,—literally, unrolling or unfolding. It is difficult to give a clear definition that will apply to each of the various theories that are held. Theories differ vastly in the extent of their application, as held by their various advocates, resulting in great confusion of terms:—

1. The atheists believe that there is no God. Hence, matter was not created, but was eternal, or came by chance. Only a mere handful of the whole human race have ever yet believed such an untenable doctrine. The existence of a Creator, is doubted or denied by extreme atheistic evolutionists, who would dethrone God, "exalt the monkey, and degrade man."

2. The first of modern scientific men to adopt the theory that all plants and animals, including man, are developed from certain original simple germs, was Lamarck, a French naturalist, in 1809. He conceded that God created matter,—nothing more. He believed in spontaneous generation, which scientific investigation has utterly disproved.

3. Darwin goes a step further and concedes there may have been a Creator of matter, and of one, or at most, a few germs, from which all vegetation and all animals came by evolution,—all orders, classes, families, genera, species, and varieties. He differs from Lamarck, by allowing the creation of one germ, possibly a few more. He says in his "Origin of Species," "I believe that animals are descended from at most only four or five progenitors; and plants from an equal or lesser number.... Analogy would lead me one step further, namely, to the belief that all animals and plants are descended from one prototype.... All the organic beings, which have ever lived on the earth, may be descended from some *one* primordial form." Darwin, because of his great scholarship, fairness, and candor, won for his theory more favor than it inherently deserves. Darwin taught that, "The lower impulses of vegetable life pass, by insensible gradations, into the instinct of animals and the higher intelligence of man," without purpose or design. None of these three hypotheses can admit the creation of man.

4. Other evolutionists, believing in the evolution of both plants and animals, nevertheless refuse to believe in the evolution of man—the most baneful application of the whole theory. Even if there were convincing proof

of the evolution of plants and animals from one germ, there is no real proof of the evolution of man. To prove this is the chief purpose of this book.

5. A fifth theory of evolution is held by many. It is called polyphyletic evolution, which means that God created numerous stocks, or beginnings of both plant and animal life, which were subject to change and growth, deterioration and development, according to his plan and purpose. So much of evolution in this sense as can be proved, is in harmony with the Bible account of the creation of plants, animals and man. The false theory of evolution is called the monophyletic, which teaches that all species of plants and animals including man, developed from one cell or germ which came by creation or spontaneous generation. Evolution is used throughout this book in this latter sense, unless otherwise indicated by the context. God does not create by evolution, for it can only develop what already exists.

This book is divided into three parts: In Part One, material evolution, especially the evolution of the human body, is disproved. In Part Two, the alleged proofs of evolution are examined and refuted. In Part Three, the evolution of the soul is shown to be impossible.

There are in all fifty numbered arguments, including answers to the arguments of evolutionists.

Part One
The Evolution of the Human Body
Mathematically Disproved

Any scientific theory or hypothesis must be proved first possible, then probable, then certain. To be a possible theory, it must be reconcilable with many facts; to be a probable theory, it must be reconcilable with many more; to be a certain and proven theory, it must be reconcilable with *all* the facts. Whenever it is irreconcilable with *any* fact, it should be rejected, as it can not be a true theory. Every true theory passes through these three stages,— possibility, probability, and certainty. A theory is not science, until it is certainly true, and so becomes knowledge. The evolution of man from the brute is in the throes of a desperate struggle to show that it may possibly be a true theory or hypothesis. Yet some who are ready to admit that they are "scientists," claim evolution a proven theory.

If it can be shown possible for man to have descended or ascended from the lower animals, it will require enormous additional evidence to show that such descent is probable; and still much more to make it certain.

Every scientific theory, proposed as *possible*, is reconcilable with some facts. Otherwise, it would not have been considered for a moment. Many false hypotheses have been proposed, and accepted as possible and even probable, because reconcilable with some facts. The Ptolemaic theory of the universe, making the earth the centre, around which the heavens revolved in great concentric spheres, was accepted for 1400 years from A. D. 140, because it explained many things. It corresponded with appearances. It appealed to all. Its advocates had great difficulty in reconciling it with the motions of the planets, which were therefore called planets or "wanderers." But in time the Copernican theory prevailed, because it was reconcilable with all the facts. The evidence is so abundant that all claim it the true theory. It is science. It is knowledge.

Because the Copernican hypothesis, the true theory of the universe, was opposed and rejected, it does not follow that the evolution of man is true because it is likewise opposed and rejected. If this new theory, hypothesis, or guess stands, it can only do so, because it harmonizes with all the facts.

The law of gravitation, and every other proven theory harmonizes with all the facts and with all other true theories.

It will be shown in this book, that a large number of facts can not be reconciled with evolution, especially the evolution of man, thus proving that it can not be a true theory. We really have a right to demand the proof of a theory, and to refuse consent until proved. While we are under no obligation to *disprove* an unproven theory, yet it is the shortest way to settle the matter once for all, before it has led multitudes more astray, and wrecked the faith and hopes of the young.

Prof. H. H. Newman, in his "Readings in Evolution," p. 57, says, "Reluctant as we may be to admit it, honesty compels the evolutionist to admit that there is no absolute proof of organic evolution." "If all the facts are in accord with it, and none are found that are incapable of being reconciled with it, a working hypothesis is said to have been advanced to a proven theory." Note this admission by a leading evolutionist.

Even if it should ever be proved that all plant and animal life came by evolution from one primordial germ, it would not follow that either the body or the soul of man came by evolution. All the arguments against evolution in general are valid against the evolution of man. In addition, there are many other arguments, that prove the evolution of man impossible, even if the evolution of plants and animals should ever be proved possible.

In this volume, the claim is made that the evolution of man is irreconcilable with a large number of facts. If investigation proves that we have erred in any statement of facts, or if our reasoning in any one argument or more is fallacious, we will not lose our case, as long as evolution remains irreconcilable with any other single fact. If every argument in this book were invalid, save one, that one valid argument would overthrow evolution, since every true theory must be reconcilable with all the facts. One irreconcilable fact is sufficient to overthrow evolution. And there are many!

THE UNITY OF THE HUMAN RACE

The evolution of man is not only a guess, but a very wild one; and it is totally unsupported by any convincing arguments. It can be mathematically demonstrated to be an impossible theory. Every proof of the unity of the human race in the days of Adam or Noah shatters the theory of the evolution of man. If the evolution of the human race be true, there must have been, hundreds of thousands of years ago, a great multitude of heads of the race, in many parts of the earth, without one common language or religion. The present population of the globe proves that mankind must have descended from one pair who lived not earlier than the time of Noah. The unity of languages also proves one common head about the same time. Certain beliefs and customs, common to various religions, point to one original God-given religion in historic time, in contrast to the evolution idea of many religions invented by ape-men in millions of years. The history of the world and the migration of nations point to one locality where the human race began in times not more remote, and show that man was created in a civilized state, and, therefore, never was a brute. If evolution were true, there would have been many billion times as many human beings as now exist, a great multitude of invented languages with little or no similarity, a vast number of invented religions with little, if anything, in common. Even the sciences invented and exploited by evolutionists, the Mendelian Inheritance Law and Biometry, also prove evolution impossible.

The unity of mankind is also conclusively shown by the fact that all races interbreed, the most certain test of every species.

All these facts pointing to the unity of the race in the days of Noah and of Adam are irreconcilable with the theory of evolution which denies that unity within the last two million years.

We shall present these arguments more in detail. The arguments immediately following, especially the first eight, show the unity of the human race in the days of Noah, and thus present insuperable objections to evolution, and confirm the story of man's creation and his destruction by the flood. The following is the first of fifty Arguments against the evolution of man.

1
THE POPULATION OF THE WORLD

The population of the world, based upon the Berlin census reports of 1922, was found to be 1,804,187,000. The human race must double itself 30.75 times to make this number. This result may be approximately ascertained by the following computation:—

At the beginning of the first period of doubling there would just be two human beings; the second, 4; the third, 8; the fourth, 16; the tenth, 1024; the twentieth, 1,048,576, the thirtieth, 1,073,741,824; and the thirty-first, 2,147,483,648. In other words, if we raise two to the thirtieth power, we have 1,073,741,824; or to the thirty-first power, 2,147,483,648. Therefore, it is evident even to the school boy, that, to have the present population of the globe, the net population must be doubled more than thirty times and less than thirty-one times. By logarithms, we find it to be 30.75 times. After all allowances are made for natural deaths, wars, catastrophes, and losses of all kinds, if the human race would double its numbers 30.75 times, we would have the present population of the globe.

Now, according to the chronology of Hales, based on the Septuagint text, 5077 years have elapsed since the flood, and 5177 years since the ancestors of mankind numbered only two, Noah and his wife. By dividing 5177 by 30.75, we find it requires an average of 168.3 years for the human race to double its numbers, in order to make the present population. This is a reasonable average length of time.

Moreover, it is singularly confirmed by the number of Jews, or descendants of Jacob. According to Hales, 3850 years have passed since the marriage of Jacob. By the same method of calculation as above, the Jews, who, according to the Jewish yearbook for 1922, number 15,393,815, must have doubled their numbers 23.8758 times, or once every 161.251 years. The whole human race, therefore, on an average has doubled its numbers every 168.3 years; and the Jews, every 161.251 years. What a marvelous agreement! We would not expect the figures to be exactly the same nor be greatly surprised if one period were twice the other. But their correspondence singularly corroborates the age of the human race and of the Jewish people, as gleaned from the word of God by the most proficient chronologists. If the human race is 2,000,000 years old, the period of doubling would be 65,040 years, or 402 times that of the Jews, which, of course, is unthinkable.

While the period of doubling may vary slightly in different ages, yet there are few things so stable and certain as general average, where large numbers and many years are considered, as in the present case. No life insurance company, acting on general average statistics, ever failed on that account. The Jews and the whole human race have lived together the same thirty-eight centuries with very little intermarriage, and are affected by similar advantages and disadvantages, making the comparison remarkably fair.

Also, the 25,000,000 descendants of Abraham must have doubled their numbers every 162.275 years, during the 3,988 years since the birth of his son Ishmael. These periods of doubling which tally so closely, 168.3 years for the whole race, 161.251 for the Jews, and 162.275 years for the descendants of Abraham, cannot be a mere coincidence, but are a demonstration against the great age of man required by evolution, and in favor of the 5,177 years since Noah. None of the other various chronologies would make any material difference in these calculations. The correspondence of these figures, 168.3, 161.251 and 162.275 is so remarkable that it must bring the conviction to every serious student that the flood destroyed mankind and Noah became the head of the race.

Now the evolutionists claim that the human race is 2,000,000 years old. There is no good reason for believing that, during all these years the developing dominant species would not increase as rapidly as the Jews, or the human race in historic times, especially since the restraints of civilization and marriage did not exist. But let us generously suppose that these remote ancestors, beginning with one pair, doubled their numbers in 1612.51 years, one-tenth as rapidly as the Jews, or 1240 times in 2,000,000 years. If we raise 2 to the 1240th power, the result is 18,932,139,737,991 with 360 figures following. The population of the world, therefore, would have been 18,932,139,737,991 decillion, decillion, decillion, decillion, decillion, decillion, decillion, decillion, decillion, decillion; or 18,932,139,737,991 vigintillion, vigintillion, vigintillion, vigintillion, vigintillion, vigintillion.

Or, let us suppose that man, the dominant species, originated from a single pair, only 100,000 years ago, the shortest period suggested by any evolutionist (and much too short for evolution) and that the population doubled in 1612.51 years, one-tenth the Jewish rate of net increase, a most generous estimate. The present population of the globe should be 4,660,210,253,138,204,300 or 2,527,570,733 for every man, woman and child!

In these calculations, we have made greater allowances than any self-respecting evolutionist could ask without blushing. And yet withal, it is as clear as the light of day that the ancestors of man could not possibly have lived 2,000,000 or 1,000,000 or 100,000 years ago, or even 10,000 years ago; for if the population had increased at the Jewish rate for 10,000 years, it would be more than two billion times as great as it is. No guess that ever was made, or ever can be made, much in excess of 5177 years, can possibly stand as the age of man. The evolutionist cannot sidestep this argument by a new guess. Q. E. D.

All these computations have been made upon the supposition that the human race sprang from one pair. If from many in the distant past, as the evolutionists assert, these bewildering figures must be enormously increased.

Yet we are gravely told that evolution is "science". It is the wildest guess ever made to support an impossible theory.

That their guesses can not possibly be correct, is proven also by approaching the subject from another angle. If the human race is 2,000,000 years old, and must double its numbers 30.75 times to make the present population, it is plain that each period for doubling would be 65,040 years, since $\{2,000,000/30.75\} = 65,040$. At that rate, there would be fewer than four Jews! If we suppose the race to have sprung from one pair 100,000 years ago, it would take 3252 years to double the population. At this rate, there would be five Jews!

Do we need any other demonstration that the evolution of man is an absurdity and an impossibility? If the evolutionists endeavor to show that man *may* have descended from the brute, the population of the world conclusively shows that MAN CERTAINLY DID NOT DESCEND FROM THE BRUTE. If they ever succeed in showing that all species of animals *may* have been derived from one primordial germ, it is impossible that man so came. He was created as the Bible declares, by the Almighty Power of God.

The testimony of all the experts in the famous Scopes trial in Tennessee (who escaped cross-examination) was to the effect that evolution was in harmony with *some* facts and therefore *possibly true*. The above mathematical calculations prove that the evolution of man was certainly not true. They fail to make their case even if we grant their claims. These figures prove the Bible story, and scrap every guess of the great age and the brute origin of man. It will be observed that the above calculations point to the unity of the race in the days of Noah, 5177 years ago, rather than in the days of Adam 7333 years ago, according to Hales' chronology. If the race increased at the

Jewish rate, not over 16,384 perished by the Flood, fewer than by many a modern catastrophe. This most merciful providence of God started the race anew with a righteous head.

Now, if there had been no flood to destroy the human race, then the descendants of Adam, in the 7333 years, would have been 16,384 times the 1,804,187,000, or 29,559,799,808,000; or computed at the Jewish rate of net increase for 7333 years since Adam, the population would have been still greater, or 35,184,372,088,832. These calculations are in perfect accord with the Scripture story of the special creation of man, and the destruction of the race by a flood. Had it not been for the flood, the earth could not have sustained the descendants of Adam. Is not this a demonstration, decisive and final?

2
THE UNITY OF LANGUAGES

The unity of the languages of the world proves the recent common origin of man. Prof. Max Muller, and other renowned linguists, declared that all languages are derived from one. This is abundantly proven by the similarity of roots and words, the grammatical construction and accidents, the correspondence in the order of their alphabets, etc. The words for father and mother similar in form, for example, are found in many languages in all the five great groups, the Aryan, the Semitic, the Hamitic the Turanian and Chinese groups, showing a common original language and proving the early existence of the home and civilization. The similarity of these and many other words in all of the great Aryan or Indo-European family of languages, spoken in all continents is common knowledge. Lord Avebury names 85 Hamitic languages in Africa in which the names of father and mother are similar; 29 non-Aryan languages in Asia and Europe, including Turkish, Thibetan, and many of the Turanian and Chinese groups; 5 in New Zealand and other Islands; 8 in Australia; and 20 spoken by American Indians. The French, Italian, Spanish and Portuguese are daughters of the Latin; Latin is a daughter of the Aryan; and the Aryan, together with the other sister languages is, no doubt, the daughter of the original language spoken by Noah and his immediate descendants. There can not well be more than 4 generations of languages, and the time since Noah is sufficient for the development of the 1000 languages and dialects. The American Indians have developed about 200 in 3,000 or 4,000 years. The life of a language roughly speaking, seems to range from 1000 to 3,000 years. The time since Noah is sufficient for the development of all the languages of the world. But if man has existed for 2,000,000 or 1,000,000 years, with a brain capacity ranging from 96% to normal, there would have been multiplied thousands of languages bearing little or no resemblance. There is not a trace of all these languages. They were never spoken because no one lived to speak them.

Many linguists insist that the original language of mankind consisted of a few short words, possibly not over 200, since many now use only about

300. The Hebrew has only about 500 root words of 3 letters; the stagnant Chinese, 450; the Sanscrit, about the same. All the Semitic languages have tri-literal roots. As the tendency of all languages is to grow in the number and length of words, these consisting of a few small words must have been close to the original mother tongue. No language could have come down from the great antiquity required by evolution and have so few words. Johnson's Eng. Dictionary had 58,000 words; modern Dictionaries over 300,000. The evidence points to the origin and unity of languages in the days of Noah, and proves the great antiquity of man an impossibility and his evolution a pitiful absurdity.

3
RELIGIONS

The unity of ancient religions proves the creation of man who received a divine revelation. According to evolution, all religions were evolved or invented by humanoids. In that case, we would expect them to be widely divergent; and we would be surprised, if they agreed on great and important points, and especially on points which could not be clearly arrived at by reason. For instance, what in reason teaches us that an animal sacrifice is a proper way to worship God? How could unassisted reason ever arrive at the conclusion that God is properly worshipped by sacrificing a sheep or an ox? If we grant that one section of the anthropoid host might have stumbled on the idea, how can we account for its prevalence or its universality? A very high authority says, "Sacrifices were common to all nations of antiquity, and therefore, traced by some to a personal revelation." By revelation, we learn that the animal sacrifice prefigured the Lamb slain on Calvary. It was revealed. No race of monkey-men could ever have invented the idea.

The most ancient nations worshipped God by sacrifices. Homer's Iliad (1000 B. C.) and other works of Grecian poets are full of it. All the classics, Greek and Latin, are crowded with accounts of offerings. The earliest records of the Egyptians, Babylonians, Assyrians, Hindus and Chinese speak of sacrifices long in vogue. This unity of religions on the point of animal sacrifices bespeaks revelation and not evolution.

The division of time into weeks of 7 days, prevalent among the ancients, suggests an ancient revelation in commemoration of creation as against evolution, which denies creation. The following statements from Dr. J. R. Dummelow, an eminent commentator, show that the Babylonians both divided time into weeks, and offered sacrifices, pointing to the unity of religions. "The Babylonians observed the 7th, 14th, 21st and 28th of each lunar month as days when men were subjected to certain restrictions; the king was not to eat food prepared by fire, *nor offer sacrifice*, nor consult an oracle, nor invoke curses on his enemies." They also observed the 19th of each month. It was customary, therefore, in the days of Abraham, for the Babylonians to offer sacrifices and to observe the 7th day as especially sacred. This can only be accounted for upon the assumption, that God had revealed to the human race that creation occupied 6 days or periods, and the 7th was to be observed,—all of which was doubtless handed down by tradition. There were priests and temples in the most ancient empire known.

Dr. Dummelow says: "It is now widely admitted that the Genesis account of creation contains elements of belief which existed perhaps thousands of years before the book of Genesis was written, among the peoples of Babylonia and Assyria." Many of the primeval revelations were handed down by tradition. God communed with Adam. There are many relics of the original religion: the division of time into weeks, and the institution of the Sabbath day; the sacrifices so common in the ancient religions; the general existence of priests and temples in all ages, and among all nations; marriage, the divinely authorized pillar of society; the early institution of the family, and the use of the root words for father and mother, in all the most ancient languages, and families of languages, as well as in the scattered languages of the earth spoken by the most savage. The belief in the immortality of the soul, is well nigh universal, even among tribes, who, unlike Plato, possess no power to reason it from the light of nature. In contrast, we behold the sorry spectacle of the anthropoid evolutionists of our day trying to drive from the hearts of men the hope of immortality by their "science falsely so-called." The burial of the dead is, no doubt, a relic, since animals, even of the monkey tribe, do not bury their dead.

4
PLACE OF THE ORIGIN OF MAN

The unity of the human race is further proved by the fact that it originated in one locality and not in many. The locality is the one described by Moses. And the fact that Moses correctly located the beginning of the race, when he himself had no personal knowledge, proves that he was inspired and taught of God. He never could have guessed the spot to which history and the migration of nations point, and which the evolutionists themselves are obliged to concede.

The habitable countries of the world total 50,670,837 sq. mi. We are making a generous estimate, when we suppose the garden of Eden to have been 100 mi. wide and 125 mi. long,—12,500 sq. mi. There are 4005 such areas in the habitable globe. It is located in Mesopotamia on the Tigris and Euphrates rivers.

Maps of ancient nations show that mankind radiated from this centre. The great nations of antiquity were clustered about it. The beginning of the race after the flood was in the same general locality.

Ridpath in his great history of the world, graphically shows the migrations of races and nations. With this, even evolutionists agree. They draw a line "according to Giddings," running through western Asia, in the region of the garden of Eden. Since there are 4005 such areas in the habitable globe, Moses had only one chance out of 4005 to guess the spot, if he had not been inspired of God. Anyone guessing, might have located the origin of man in any of the countries of Europe, Asia or Africa. This clearly demonstrates that God revealed the truth to Moses, and that the story of creation is true and of evolution false.

If evolution were true, there must have been, 6,000 years ago, many heads to the race, in many places. It is incredible that there would be but one spot where brutes became humans. There would be an innumerable host of anthropoid brutes, in many parts of the world, in all gradations. Who can believe that one species or one pair forged ahead so far as to become human?

5
CIVILIZATIONS

The early civilization of man points to his creation, not his evolution. Evolution requires many centers of civilization; creation, only one. Of course, if man is descended from an ancient ape-like form, and from the Primates and their brute progeny, he must have been as uncivilized and brutish as any baboon or gorilla today, or the apes, which, last year, horribly mangled the children at Sierra Leone. He must have worked his way up into civilization. The records, as far back as they go, prove that the original condition of man was a state of civilization, not savagery. Man fell down, not up.

The recent explorations in the tomb of Tutankhamen, in Egypt, and the more recent explorations of the tomb of a still more ancient Egyptian monarch, show that a high degree of civilization prevailed from 2000 to 1300 B.C. The art displayed in the carvings and paintings, and the skill of the artisans are beyond praise. They had knowledge even of what are now lost arts. They had a written language 300 years before Homer wrote his immortal Iliad. Yet many higher critics claim that writing was unknown in the days of Moses and Homer. They declare that the Iliad, a poem in 24 books, was committed to memory, and handed down from generation to generation, 400 years with all its fine poetic touches. Monstrous alternative! Indeed we are even told that "Many men must have served as authors and improvers." The mob of reciters improved the great epic of Homer! Scarcely less brilliant is the suggestion of another higher critic that, "Homer's Iliad was not composed by Homer, but by another man of the same name"!

The laws of Hammurabi, who is identified as the Amraphel of Scripture, Gen. 14:1, and who was contemporary with Abraham, were in existence many hundred years before Moses, and showed a high state of civilization, which began many hundred years before Abraham. The literature of China goes back to 2000 B. C. The earliest civilization of China, Egypt, Assyria, and Babylonia, reaching to 2500 B.C., or earlier, points to a still earlier civilization, which likely reaches back to the origin of the human race.

It is admitted that the earliest (Sumerian) civilization began on the Euphrates, near the garden cif Eden. They had temples and priests, and, therefore, religion prevailed as well as civilization. The first great empires clustered around the places where Adam and Noah lived. No other civilization recorded in any quarter reaches farther back.

We quote from the New International Encyclopedia: "The Sumerian language is probably the oldest known language in the world. From the Sumerian vocabulary, it is evident that the people who spoke this language had reached a comparatively high civilization."

The monuments show that in early historical times, man was in a state of civilization. There are no monuments of man's civilization prior to historical time.

Higher critics have said that Moses could not have written the Pentateuch because writing was unknown in his day. Yet Prof. A. H. Sayce, D.D., LL.D., of Oxford University, one of the greatest archaeologists the world ever knew, writes: "Egypt was the first to deliver up its dead. Under an almost rainless sky, where frost is unknown, and the sand seals up all that is entrusted to its keeping, nothing perishes except by the hand of man. The fragile papyrus, inscribed it may be 5,000 years ago, is as fresh and legible as when its first possessor died.

"In Egypt, as far back as the monuments carry us, we find a highly-developed art, a highly organized government, and a highly-educated people. Books were multiplied, and if we can trust the translation of the Proverbs of Ptah-hotep, the oldest existing book in the world, there were competitive examinations, [civil service!] already in the age of the sixth Egyptian Dynasty.... We have long known that the use of writing for literary purposes is immensely old in both Egypt and Babylonia. Egypt was emphatically a land of scribes and readers. Already in the days of the Old Empire, the Egyptian hieroglyphs had developed into a cursive hand."

From the Tel el-Amarna tablets, discovered in Upper Egypt, we know that for 100 years people were corresponding with each other, in the language of Babylonia in cuneiform characters. Libraries existed then, and "Canaan in the Mosaic age, was fully as literary as was Europe in the time of the Renaissance." Ancient Babylonian monuments testify to the existence of an ancient literary culture. The results of the excavations by the American Expedition, published by Prof. Hilprecht, of the U. of Pa., show that in the time of King Sargon of Accad, art and literature flourished in Chaldea. The region of the garden of Eden was the pivot of the civilization of the world. From this region radiated the early civilization of Babylonia, Assyria and Egypt. And the advanced degree implies centuries of prior civilization. The origin of man and the earliest civilization occurred in the same region. Ur explorations (1927) show high art, 3000 B.C.

The earliest records show man was civilized. He lived in houses, cities and towns, read and wrote, and engaged in commerce and industry. To be sure, he did not have the inventions of modern times. If all these were necessary, then there was no civilization prior to the 20th century. Prof. J. Arthur Thompson, of Aberdeen, an evolutionist, says: "Modern research is leading us away from the picture of primitive man as brutish, dull, lascivious and bellicose. There is more justification for regarding primitive man as clever, kindly, adventurous and inventive."

It is admitted that cannibalism was not primeval. The two great revolting crimes of barbarism, cannibalism and human sacrifices, only prevailed when man had fallen to the lowest depths, not when he had risen out of savagery to the heights. The assertion that man was originally a brute, savage and uncivilized is pure fiction, unsupported by the facts. The original civilization of mankind supports the Bible, and upsets evolution.

6
THE MENDELIAN INHERITANCE LAW

The unity of the human race is further established by Mendel's Inheritance Discovery on which evolutionists so much rely. G. Mendel, an experimenter, found that when he crossed a giant variety of peas with a dwarf variety, the off-spring were all tall. The giants were called "dominant"; the disappearing dwarfs, "recessive". But among the second generation of this giant offspring, giants and dwarfs appeared in the proportion of 3 to 1. But when these dwarfs were self-fertilized, successive generations were *all* dwarfs. The recessive character was not lost, but appeared again. Experiments with flowers likewise show that the recessive color will reappear.

Also experiments with the interbreeding of animals have shown similar results. The recessive or disappearing characteristics, or the disappearing variety, will appear again, in some subsequent generation, and sometimes becomes permanent. This law prevails widely in nature, and the recessive traits appear with the dominant traits. "If rose-combed fowl were mated with single-combed fowl, the offspring were all rose-combed, but when these rose-combed fowl were mated, the offspring were again rose-combed and single-combed.... If gray rabbits were mated with black rabbits, their hybrids were all gray, the black seemingly disappearing, but when the second generation were mated, the progeny were again grays and blacks." — God or Gorilla — p. 278. *The recessive character always reappears.*

Apply these widely prevalent laws to dominant man and his recessive alleged brute ancestor. The simian characteristics would appear in some generations, if not in many. We would expect many offspring *to have the recessive character of the ape*, and we ought not to be surprised, if some recessive stock became permanent.

Following analogy, we ought to look for a tribe of human beings that had degenerated into apes. That we find no such recessive characteristics even among the most degenerate savages, and no such ape-like tribe of human beings, is a decisive proof that man never descended from the brute. Else such recessive characteristics, according to the Mendelian Law, would be sure to appear. We would also find monkeys and apes, — the recessive species — descended from man.

7
BIOMETRY

Even new sciences, founded by evolutionists, bear witness against their theory. Mendel's Inheritance Law is one, as we have seen; Biometry is another. It was proposed and advocated by Sir Francis Galton, a cousin of Charles Darwin. He expected it to be a great prop to evolution; on the other hand, it is another proof of the unity of our race in Noah's day, and hence fatal to their theory. Biometry is defined to be the "statistical study of variation and heredity." It bears heavily against the great age of man.

One of the leading exponents of Biometry, Dr. C.B. Davenport, Secretary of the Eugenics section of the American Breeders' Association concludes that "No people of English descent are more distantly related than thirtieth cousin, while most people are more nearly related than that." Professor Conklin, of Princeton University, approves this conclusion, and adds, "As a matter of fact most persons of the same race are much more closely related than this, and certainly *we need not go back to Adam nor even to Shem, Ham or Japheth to find our common ancestor.*" Dr. Davenport, therefore, says that the English may find a common ancestor thirty-two generations ago; Professor Conklin admits that we need not go further back than Noah to find a common ancestor of all mankind. Noah, therefore, must have been the head of the race. Evolutionists admit we need go no farther back than Noah to find the head of the race, and the population, as we have seen, proves the same thing, and disproves every guess they have made of the great age of man. We have descended from Noah and not from the brute.

This same Professor Conklin says that our race began 2,000,000 years ago (60,000 generations). How is it possible that we must go back sixty thousand generations for a common ancestor, when thirty-two generations will suffice for the English, and about 200 generations since Noah, for the whole race? If we, by the laws of biometry, can find a common ancestor in Noah, we can not possibly go back 2,000,000 years to find one. Professor Conklin's admission refutes his claim of 2,000,000 years for man. Biometry proves that age absolutely impossible.

If the progeny of this ape-like ancestor inter-bred for many generations,— as certainly would have been the case—then we are not only descended from all the monkey family, the baboon, gorilla, ape, chimpanzee, orangutang lemur (H. G. Wells' ancestor), mongoose, etc., but are also related to all their progeny. Glorious ancestors! In our veins runs the blood of them all, as well as the blood of the most disgusting reptiles. And yet Professor H. H. Newman, an eminent evolutionist, in a letter to the writer, says, "The evolution idea is an ennobling one."! But biometry saves us from such repulsive forbears, by proving it could not be so.

Biometrists find that there is a Law of Filial Regression, or a tendency to the normal in every species, checking the accumulation of departures from the average, and forbidding the formation of new species by inheritance of peculiarities. The whole tendency of the laws of nature is against the formation of new species, so essential to evolution. The species brings forth still "after its kind." "On the average, extreme peculiarities of parents are less extreme in children." "The stature of adult offspring must, on the whole, be more mediocre than the stature of the parents." Gifted parents rarely have children as highly gifted as themselves.

The tendency is to revert to the normal in body and mind. Nature discourages the formation of new species, evolutionists to the contrary notwithstanding. "Like produces like" is a universal and unchangeable law. God has forbidden species to pass their boundaries; and, if any individual seems to threaten to do so, by possessing abnormal peculiarities, these are soon corrected, often in the next generation. Even Professor H. H. Newman says, "On the whole, the contributions of biometry to our understanding of the causes of evolution are rather disappointing." A science that upsets evolution is certainly disappointing to evolutionists.

8
NO NEW SPECIES NOW

They tell us that 3,000,000 species of plants and animals developed from one primordial germ, in 60,000,000 years. How many new species should have arisen in the last 6,000 years? Now 20 doublings of the first species of animals would make 1,048,576 species, since 2 raised to the 20th power becomes 1,048,576. Again we will favor the evolutionists, by omitting from the calculation all species of animals in excess of 1,048,576. Therefore, on an average, each of the 20 doublings would take 1/20 of 60,000,000 years, or 3,000,000 years; and, therefore, 1/2 of the entire 1,048,576 species, or 524,288 species, must have originated within the last 3,000,000 years. Can that be the case? Certainly not.

And since the number of species must have increased in a geometrical ratio, 2097 species must have arisen or matured within the last 6000 years—an average of one new species of animals every 3 years. How many species actually have arisen within the last 6000 years? 2000? 200? or 2? It is not proven that *a single new species has arisen in that time*. Not one can be named. If approximately 2000 new species have not arisen in the last 6000 years, the evolution of species can not possibly be true. Even Darwin says: "In spite of all the efforts of trained observers, not one change of species into another is on record." Sir William Dawson, the great Canadian geologist, says: "*No case is certainly known in human experience* where any species of animal or plant has been so changed as to assume all the characteristics of a new species."

Indeed, a high authority says: "Though, since the human race began, all sorts of artificial agencies have been employed, and though there has been the closest scrutiny, yet *not a distinctively new type of plant or animal*, on what is called broad lines, has come into existence."

Not a single new species has arisen in the last 6000 years when the theory requires over 2000. Evolutionists admit this. Prof. Vernon Kellogg, of Leland Stanford University, in his "Darwinism of Today," p. 18, says:—"Speaking by and large, we only tell the general truth when we declare that no indubitable cases of species forming, or transforming, that is, of descent, have been observed.... For my part, it seems better to go back to the old and safe *ignoramus standpoint*."

Prof. H. H. Newman, of Chicago University, in answer to the writer's question, "How many new species have arisen in the last 6000 years?" wrote this evasive reply: "I do not know how to answer your questions.... None of us know just what a species is. [If so, how could 3,000,000 species be counted, the number, he says, exists?].... It is difficult to say just when a new species has arisen from an old." He does not seem to know of a single new species within the last 6,000 years.

The same question was asked of Dr. Osborn, of Columbia University, N. Y. The answer by R. C. Murphy, assistant, was equally indefinite. He wrote: "From every point of view, your short note of Aug. 22nd raises questions, which no scientific man can possibly answer. We have very little knowledge as to just when any particular species of animal arose." In a later letter, he says: "I have no idea whether the number of species which have arisen during the last 6000 years is 1 or 100,000."

Should those who "do not know" speak so confidently in favor of evolution, or take the "old and safe *ignoramus*" standpoint, as Prof. Kellogg suggests?

The number of existing species can not be explained upon the ground of evolution, but only upon the ground of the creation of numerous heads of animal and plant life, as the Scriptures declare.

We have a right to increase the pressure of the argument, by introducing into the calculation, the total of 3,000,000 species of plants and animals which would require 6355 new species within the last 6000 years, or an average of more than one new species a year! And they can not point to one new species in 6000 years, as they confess. Dr. J. B. Warren, of the University of California, said recently: "If the theory of evolution be true, then, during many thousands of years, covered in whole or in part by present human knowledge, there would certainly be known at least a few instances of the evolution of one species from another. *No such instance is known*."

Prof. Owen declares, "No instance of change of one species into another has ever been recorded by man."

Prof. William Bateson, the distinguished English biologist, said, "It is impossible for scientists longer to agree with Darwin's theory of the origin of species. No explanation whatever has been offered to account for the fact that, after forty years, no evidence has been discovered to verify his genesis of species."

Although scientists have so largely discarded Darwin's theory, the utter lack of new species in historic time, when so many are required by *every* theory of evolution, is a mathematical demonstration that the whole theory of evolution must be abandoned. Q. E. D. Why do they still insist it *may be true*?

9
MATHEMATICAL PROBABILITY

Mathematical Probability is a branch or division of mathematics by means of which the odds in favor or against the occurrence of any event may be definitely computed, and the measure of the probability or improbability exactly determined. Its conclusions approximate certainty and reveal how wild the guesses of evolutionists are.

The evolution of species violates the rule of mathematical probability. It is so improbable that one and only one species out of 3,000,000 should develop into man, that it certainly was not the case. All had the same start, many had similar environments. Yet witness the motly products of evolution: Man, ape, elephant, skunk, scorpion, lizard, lark, toad, lobster, louse, flea, amoeba, hookworm, and countless microscopic animals; also, the palm, lily, melon, maize, mushroom, thistle, cactus, microscopic bacilli, etc. All developed from one germ, all in some way related. Mark well the difference in size between the elephant, louse, and microscopic hookworm, and the difference in intellect between man and the lobster!

While all had the same start, only one species out of 3,000,000 reached the physical and intellectual and moral status of man. Why only one? Why do we not find beings equal or similar to man, developed from the cunning fox, the faithful dog, the innocent sheep, or the hog, one of the most social of all animals? Or still more from the many species of the talented monkey family? Out of 3,000,000 chances, is it not likely that more than one species would attain the status of man?

"Romanes, a disciple of Darwin, after collecting the manifestations of intelligent reasoning from every known species of the lower animals, found that they only equaled altogether the intelligence of a child 15 months old." Then man has easily 10,000,000 times as much power to reason as the animals, and easily 10,000,000,000 times as much conscience. Why have not many species filled the great gap between man and the brute? Out of 3,000,000 births, would we expect but one male? Or one female? Out of 3,000,000 deaths, would we expect all to be males but one? To be sure, all the skeletons and bones found by evolutionists belong to males except one. Strange! If 3,000,000 pennies were tossed into the air, would we expect them all to fall with heads up, save one? The Revolutionary war, out of 3,000,000 people, developed one great military chieftain, but many more approximating his ability; one or more great statesmen with all gradations down to the mediocre; scholars and writers, with others little inferior; but there was no overtowering genius 10,000,000 or 10,000,000,000 times as great as any other. We would be astonished beyond measure, if any great genius should rise in any nation as far ahead of all others, as the species of mankind is ahead of all other species. It is unthinkable that one species and only one reached the measureless distance between the monkey and man. It violates mathematical probability.

We have a right to expect, in many species and in large numbers, all gradations of animals between the monkey and man in size, intellect, and spirituality. Where are the anthropoids and their descendants alleged to have lived during the 2,000,000 years of man's evolution? They can not be found living or dead. They never existed. Creation alone explains the great gap. What signs have we that other species will ever approximate, equal or surpass man in attainments? Can we hope that, in the far distant future, a baboon will write an epic equal to Milton's Paradise Lost, or a bull-frog compose an oratorio surpassing Handel's Messiah?

We find all gradations of species in size from the largest to the smallest. Why not the same gradation in *intelligence, conscience and spirituality*? The difference in brain, capacity and intelligence between man and the ape is 50% greater than the difference in size between the elephant and the housefly. There are many thousands of species to fill the gap in size. Why not many thousands to fill the greater gap in intelligence? Evidently no species became human by growth. Many species like the amoeba, and the microscopic disease germs, have not developed at all but are the same as ever. Many other species of the lower forms of life have remained unchanged during the ages. If the tendency is to develop into the higher forms of life, why do we have so many of those lower forms which have remained stationary? Growth, development, evolution, is not, by any means, a universal rule.

Evolution is not universally true in any sense of the term. Why are not fishes *now* changing into amphibians, amphibians into reptiles, reptiles into birds and mammals, and monkeys into man? If growth, development, evolution, were the rule, there would be no lower order of animals for all have had sufficient time to develop into the highest orders. Many have remained the same; some have deteriorated.

And now we have a new amendment to the theory of evolution: We are told that the huge Saurians (reptiles) overworked the development idea, and became too large and cumbersome, and hence are now extinct. Prof. Cope says:—"Retrogression in nature is as well established as evolution." It seems that man also has, contrary to all former conceptions, reached the limit of his development, if he has not already gone too far.

Prof. R. S. Lull says, (Readings p. 95) "Man's physical evolution has virtually ceased, but in so far as any change is being effected, it is largely retrogressive. Such changes are: Reduction of hair and teeth, and of hand skill; and dulling of the senses of sight, smell and hearing upon which active creatures depend so largely for safety. That sort of charity which fosters the physically, mentally and morally feeble, and is thus contrary to the law of natural selection, must also, in the long run, have an adverse effect upon the race." Too bad that Christian charity takes care of the feeble, endangering evolution, and the doctrine that the weak have no rights that the strong are bound to respect! We are not surprised that Nietzsche, whose insane philosophy that *might is right*, helped to bring on the world war, died in an insane asylum.

After all, evolution is not progress and development, but retrogression and deterioration as well.

But evolutionists, compelled by the requirements of their theory, have added another amendment, which will seem ridiculous to some:

Environment has had an evolution as well as plants and animals! Having denied the existence of God, or his active control and interference, they must account for environment by evolution. Listen:—"Henderson points out that environment, no less than organisms, has had an evolution. Water, for example, has a dozen unique properties that condition life. Carbon dioxide is absolutely necessary to life. The properties of the ocean are so beautifully adjusted to life that we marvel at the exactness of its fitness. [Yet no design!]. Finally, the chemical properties of carbon, hydrogen and oxygen are equally unique and unreplaceable. The evolution of environment and the evolution of organisms have gone hand in hand." And all by blind chance! Is it not a thousand times better to believe that all things were created by an all-wise and all powerful God? How could a lifeless environment come by evolution? If we would listen to them, we would be told that the ocean, the atmosphere, heat, light, electricity, all the elements, the starry heavens, and all the universe, and religion itself, came by evolution, some grudgingly granting that God *may* have created matter in the beginning.

It is unreasonable to believe that one species and only one out of 3,000,000 by evolution should attain the status of mankind; and that one species and only one species of the primates should reach the heights of intelligence, reason, conscience and spirituality. Huxley says, "There is an enormous gulf, a divergence practically infinite, between the lowest man and the highest beast."

To declare that our species alone crossed this measureless gulf, while our nearest relatives have not even made a fair start, is an affront to the intelligence of the thoughtful student. It does fierce violence to the doctrine of mathematical probability. It could not have happened.

10
THE AGE OF THE EARTH

The estimates of the age of the world vary from 16,000,000 years to 100 times this number or 1,600,000,000 years. Even H.G. Wells admits these estimates "rest nearly always upon theoretical assumptions of the slenderest kind." This is undoubtedly true of the reckless estimates of evolutionists, whose theory requires such an enormous length of time that science can not concede it. Prof. H.H. Newman says, "The last decade has seen the demise (?) of the outworn (?) objection to evolution, based on the idea that there has not been time enough for the great changes that are believed by evolutionists to have occurred. Given 100,000,000 or 1,000,000,000 years since life began we can then allow 1,000,000 years for each important change to arise and establish itself."

An objection is not "outworn" until answered, and to speak of the demise of a generally accepted theory is hardly scientific. We will not allow the evolutionist to dismiss so weighty an objection with a wave of the hand. Prof. Newman, in his "Readings in Evolution," p. 68, gives 60,000,000 years as the probable time since life began. The writer, having based arguments upon that assumption, was surprised to receive a private letter from him claiming that life has existed for 500,000,000 years. Indeed Prof. Russell, of Princeton, says, in his "Rice Lectures," that the earth is probably 4,000,000,000 years old, possibly 8,000,000,000! We can do nothing but gasp, while the bewildering guesses come in, and we wait for the next estimate. We note their utter abandon, as they make a raid on God's eternity to support a theory that would dethrone Him. But these extravagantly long periods required by the theory, science cannot grant, for the following reasons: —

1. According to the nebular hypothesis, and Helmholtz's contraction theory, accounting for the regular supply of heat from the sun, the sun itself is not likely more than 20,000,000 years old, and, of course, the earth is much younger. Both of these theories are quite generally accepted by scientists, and have much to support them. Prof. Young, of Princeton, in his Astronomy, p. 156, says, "The solar radiation can be accounted for on the hypothesis first proposed by Helmholtz, that the sun is shrinking slowly but continually.

It is a matter of demonstration that an annual shrinkage of about 300 feet in the sun's diameter would liberate sufficient heat to keep up its radiation without any fall in its temperature".... The sun is not simply cooling, nor is its heat caused by combustion; for, "If the sun were a vast globe of solid anthracite, in less than 5,000 years, it would be burned to a cinder." We quote from Prof. Young's Astronomy: "We can only say that while no other theory yet proposed meets the conditions of the problem, this [contraction theory] appears to do so perfectly, and therefore has high probability in its favor." "No conclusion of Geometry," he continues, "is more certain than this,—that the shrinkage of the sun to its present dimensions, from a diameter larger than that of the orbit of Neptune, the remotest of the planets, *would generate about 18,000,000 times as much heat as the sun now radiates in a year*. Hence, if the sun's heat has been and still is wholly due to the contraction of its mass, it can not have been radiating heat at the present rate, on the shrinkage hypothesis, for more than 18,000,000 years; and on that hypothesis, the solar system in anything like its present condition, can not be much more than as old as that." If so, evolution, on account of lack of time, can not possibly be true. If we add many millions of years to this number, or double it more than once, the time is not yet sufficient. For if the sun is 25,000,000, or even 50,000,000 years old, by the time the planets are thrown off, in turn, from Neptune to the earth, and then the earth cooled sufficiently for animal life, only a few million years would be left for evolution, a mere fraction of the time required. This is a mathematical demonstration that evolution can not be true. The same calculations, 18,000,000 to 20,000,000 years, have been made by Lord Kelvin, Prof. Todd and other astronomers.

2. The thickness of the earth's crust is fatal to the theory of the great age of the earth, required by evolution. The temperature increases as we descend into the earth, about one degree for every 50 feet, or 100 degrees per mile. Therefore, at 2 mi., water would boil; at 18 mi., glass would melt (1850°); at 28 mi., every known substance would melt (2700°). Hence the crust is not likely more than 28 miles thick,—in many places less. Rev. O. Fisher has calculated that, if the thickness of the earth's crust is 17.5 mi., as indicated by the San Francisco earthquake, the earth is 5,262,170 years old. If the crust is 21.91 mi. thick, as others say, the age would be 8,248,380 years. Lord Kelvin, the well known scientist, who computed the sun's age at 20,000,000 years, computed the earth's age at 8,302,210 years. Subtract from these computations, the years that must have elapsed before the earth became cool enough for animal life, and the few millions of years left would be utterly insufficient to render evolution possible. Note how these figures agree with the age of the earth according to the Helmholtz contraction theory. The thinness of the earth's crust is also proven by the geysers, the

volcanoes, and the 9000 tremors and earthquakes occurring annually in all parts of the world.

3. The surface marks on the earth point to much shorter periods of time since the earth was a shoreless ocean than those required by evolutionists, who are so reckless in their guesses and estimates. They help themselves to eternity without stint. Charles Lyell, a geologist of Darwin's time, set the example when he said, "The lowest estimate of time required for the formation of the existing delta of the Mississippi is 100,000 years." According to careful examination made by gentlemen of the Coast Survey and other U.S. officers, the time was 4,400 years—a disinterested decision. In the face of these three arguments, it is a bit reckless to say the earth has existed, 1,600,000,000 years,—nearly 100 times as long as proven possible by mathematical calculation. And still more reckless is the estimate of Prof. Russell, 4,000,000,000 to 8,000,000,000 years, founded on the radio-activity theory. All these wild estimates are out of the question.

The recession of the Niagara Falls from Lake Ontario required only 7,000 to 11,000 years. It required only 8,000 years for the Mississippi River to excavate its course.

Prof. Winchell estimates that the Mississippi River, has worn a gorge 100 feet deep, 8 miles long, back to the Falls of St. Anthony, in about 8,000 years. The whole thickness of the Nile sediment, 40 feet in one place, was deposited in about 13,000 years. Calculations by Southall and others from certain strata have fixed man's first appearance on the earth at 8,000 years, in harmony with Scripture.

LeConte, in his Geology, p. 19, says, "Making due allowance for all variations, it is probable that all land-surfaces are being cut down and lowered by rain and river erosion, at a rate of one foot every 5,000 years. At this rate, if we take the mean height of lands as 1200 feet, and there be no antagonistic agency at work raising the land, all lands would be cut down to the sea level and disappear in 6,000,000 years."

May we not from these data, judge approximately of the age of the world, and show by this proof also, that the world can not be at all as old as the evolution theory demands? If the surface of the earth will be worn down 1200 feet on an average in 6,000,000 years, would it not also be true that the surface has been worn down at least 1200 feet in the last 6,000,000 years? For the higher the surface, the more rapid the erosion. And if the earth is 8,302,210 years old, as Lord Kelvin computes, then at the same rate, it must have been worn down an average of 1660 feet,—38% more than remains. Is this not a fair estimate for the amount of erosion and the age of the world? How high must the land have averaged, if the world is even 60,000,000 years old?

If this be true, how long would it have taken erosion in the past, to reduce the land to its present configuration,—the short period indicated by science, or the immensely long period required by evolution?

But the evolutionists are clinging to the radio-activity theory desperately, an S.O.S. of a lost cause, depending, like evolution, on a great many assumptions, and unproven hypotheses. The assumption is that a radio-active substance, like uranium, "decays," or passes into many other substances, of which radium is one, finally producing lead in 1,000,000,000 years or more. From this theory, Prof. Russell concludes that the earth is 4,000,000,000 to 8,000,000,000 years old, and the sun is older still. During this inconceivably long period, the sun was giving out as much heat as at present, which is 2,200,000,000 times as much as the earth receives. The heat of the sun can not be accounted for, by either the combustion or cooling off theory. By the commonly accepted contraction theory, the heat has been maintained only about 20,000,000 years. How could it have been sustained 4,000,000,000 to 8,000,000,000 years? Prof. Russell answers: "We must therefore *suppose* that energy from an 'unknown source' becomes available at exceedingly high temperatures.... We can not do more than *guess* where it is hidden." Is this scientific? This theory, moreover, is interlocked with Einstein's theory of Relativity, which holds that all energy has mass, and all mass is equivalent to energy. Although 2700 books have been written, pro and con, upon Einstein's theory, yet he says only 12 men understand it, and a scientist retorts that Einstein can not be one of the 12. The contraction theory, the thickness of the cooled crust of the earth, and the conformation of its surface, all give mathematical proof that evolution is impossible because of lack of time.

11
GEOLOGY AND HISTORY

During the historical period, the species have remained unchanged. If over 1,000,000 species of animals have arisen in the 60,000,000 years, as is claimed, over 2000 of them must have arisen in the last 6,000 years. As evolutionists can not name a single new species that has arisen within that time, their theory falls to the ground. No species in that time, has passed into another. No species has been divided into two or more. No lower species has advanced into a higher. History gives no scrap of evidence in support of evolution. Even the horse, whose history has been dubiously traced for 3,000,000 years, has been a horse unchanged for the last 6,000 years. Even if the missing links in the development of the horse *could* be supplied, it would still be the same species all the while. But there are no transitional forms showing alleged changes in the development of the horse from the four-toed creature of squirrel like size. Many varieties and individuals under the skill of man have been developed and improved, but not a single new species in historic time. There are 5,000 varieties of apples but no new species. But when the evolutionist is hard pressed to answer, he takes to the wilds of eternity where it is hard to pursue him, and to check up on his guesses. He answers that changes are so slow, and take so many millions of years, that they can not tell of a single new species in the last 6,000 years, when over 2,000 are required.

He appeals to Geology, which is history down to historic time, expecting to take advantage of the ignorance of the careless student.

But Geology will not aid him to prove his reckless theory. Even Darwin complained that the evidences from Geology were scanty. Geology testifies: The genera and species of fossil animals are as distinct as those now living; new species appear at certain epochs entirely different from those which preceded; often the most perfect specimens of a new species appear at the beginning of a geologic period rather than at its close, leaving no room for evolution; no species is shown changing into another; and many species are largest at the beginning. As Geology is brought in as a hopeful witness by evolutionists, they are bound by a well-known principle of law, to accept the statements of their own witness even though fatal to their theory.

For them, Geology furnishes sorry evidence concerning the evolution of man from the brute. The great scheme of evolution claims as its chief support four geologic "finds." We can not be certain that any one of these has the slightest evidential value. An ardent evolutionist, Dr. Dubois, found a few bones, part ape, part human, buried in the river *sands*, 40 feet deep. They were scattered 50 feet apart, no two joined together. They called this strange creature pithecanthropus, and fixed its age at 750,000 years; others reduced it to 375,000 years. These few bones are no doubt from a modern ape and modern man.

The Heidelberg Jaw was also found *in the sand,* and is guessed to be 700,000 years old. It is hard to be respectful while they gravely tell such stories. But the next is even worse: The Piltdown man, alias the Piltdown fake, fabricated out of a few bones of a man and a few of an ape. It is rejected as a fabrication even by many evolutionists.

The Neanderthal man lived, they say, about 50,000 years ago. A part of a skull was found in a cave.

All the bones purporting to belong to these four creatures would not together make one complete skeleton, or even one complete skull. A child could carry all this "evidence" in a basket. These skulls can be duplicated by abnormal skulls in many graveyards today. Scientists are not certain they belong to the same individual. Part ape, part human. A desperate effort to get convincing evidence, where there is none. We can not be certain they lived in the age claimed. Scientists, even evolutionists, differ widely.

In contrast to this scant and uncertain evidence, Ales Hrdlicka, of the Smithsonian Institution, speaking of a single locality, says, "Near Lyons, France, the skeletons of 200,000 prehistoric horses are scattered. In one cave in Moravia, there are enough mammoth teeth to fill a small sized hall.... From the Heidelberg man, there is practically no record for about 200,000 years. The kinship of the Piltdown Java and Heidelberg man *is open to dispute.* The Neanderthal man may not have been a direct ancestor, of the species which produced Shakespeare, Napoleon and Newton." Remains of the unchanged ape are abundant. But the alleged human remains are scanty and uncertain.' Now if there were millions and billions of human beings developing from the brutes, should we not expect as many remains as of horses and mammoths and apes? We do not have millions of them, simply because they did not exist. Is not this well nigh a demonstration?

Shall we, upon this scant and uncertain evidence, accept a theory that shocks the reason and the moral sense of mankind, and which leads naturally to infidelity and atheism, and takes away even our hope of immortality?

Later in this volume we will consider more fully the alleged proofs from these geologic "finds."

Prof. Charles Lyell said: "In the year 1806, the French Institute enumerated not less than 80 geological theories which were hostile to the Scriptures; but not one of these theories is held today."

Many have come to the hasty conclusion that there was a continuous elaboration or a progressive growth among all species. True in some cases, but by no means universal. Many species have remained stable for millions of years; many have retrograded and deteriorated. Indeed, some evolutionists claim man has retrograded.

Many species of animals have been larger than their modern descendants. Many species show no change. All the bacilli remain the same microscopic species, even those too microscopic to be seen or isolated. They multiply the same, and produce the same diseases. How can there be growth in the microscopic world either animal or vegetable? The doctrine that there is a development and a growth among all species of animals or plants, is contradicted by the facts. If that doctrine were true, there would be no lower order of animals after so many millions of years of growth. All would have been large and of a high order like others. Since we find a majority of all animal species less in size than the fly, there has been little growth in most species, and in many, none at all. The amoebae, one celled animals, smaller than a small pin-head, have existed unchanged since life began. If plants and animals all developed from a one-celled animal, such as the amoeba, why did not the amoeba develop? Or, if some developed, why not all? Certainly there would not remain a great multitude of species in the microscopic world.

Of many species small and large, we have many fossils preserved but *no transitional forms*. The archæopteryx, a bird with a feathered tail, is the only alleged transitional form between the reptiles and the birds. Only two specimens of this same animal have been found. This could easily be an exceptional species of created birds differing no more from the normal bird than the ostrich or humming bird. If there were transitional forms we ought to have them by the millions. No transitional forms have been found between reptiles and mammals; and we have seen that there are no reliable forms between man and mammals. The numerous missing links make a chain impossible. Evolution is not simply growth or change, but the development of all species from one germ.

12
GEOGRAPHICAL DISTRIBUTION

Geographical Distribution, another witness claimed by the evolutionists, bears testimony, which they are bound, in law, to receive.

We find animals whose power of locomotion is very limited, scattered all over the world, like the mollusca and crustacea, embracing a large number of families, genera, and species. It is incredible that these all originated in one place, and from one germ, and migrated to distant parts of the world. The oyster, for example, is found in Europe, Africa, North and South America. There are over 200 species, found in all warm tempered climates, but none in the coldest regions. How could they cross the ocean and be distributed along all continents? They are soon attached to solid rocks, or other supports, and do not move at all. And if they do, how could they cross thousands of miles of ocean barren of all food?

Dr. George W. Field, an expert authority, says the oysters of Europe are unisexual, but in America, they are double-sexed. How could one be derived from the other? Even the oyster is too much for the evolutionist. The same argument applies to a great multitude of species, that have little or no powers of locomotion.

If all plants and animals originated from one germ in one place, how can plants, indigenous to a single continent, or hemisphere, be accounted for? Why, for example, was there no maize, or Indian corn, in the old world? Or tomatoes, potatoes, or any other plants indigenous to America? If these once existed in the old world, as they must have done, according to the theory, why were they found in America alone?

Here we quote from Prof. Agassiz, one of the greatest authorities the world ever knew: "I will, therefore, consider the transmutation theory of species as a *scientific mistake, untrue in its facts, unscientific in its method, and mischievous in its tendency.*" (*Italics ours and yours*).

13
GOD NOT ABSENT NOR INACTIVE

The theory that God is absent or inactive is as untenable and God-dishonoring as the discarded theory of atheism itself.

Evolution, as held by many, harmonizes with and supports the false and impossible assumption that God created one, or at most, a few germs, from which all animal species including man, and plants developed, by "natural law." This theory seems plausible to those who do not examine it too closely. It does not deny the existence of God, and concedes he may have created one or more germs, but delegated the development of an orderly world to "natural law." Thus his activities are no longer needed. Perhaps they entertain the thought that God must grow weary under the active and sleepless control of the universe, if not of the world alone. They lose sight of the fact that a God of infinite mind and power can not be wearied by any possible complications, or any required amount of energy. Rather, the exercise of unlimited energy is a source of pleasure and happiness. May we not learn this from the boundless extent of the universe? Creation is not a task, but a great satisfaction. If God finds so much happiness in creating a boundless universe, would he renounce the pleasure of the active care and control of 3,000,000 species?

The hypothesis that God delegates to "law" the evolution of the universe, the world, and all species, is untenable, because no law, human or divine, can enforce itself. Law has no power. It is not a being, a creature, a living thing. It is absolutely helpless. It can not be God's agent to carry out his will. Why the need of it? Why should not God use his power direct to do his will? What gain in creating and employing an agent? Which would be easier, to execute his own will, or delegate it to a law?

His law is simply the record of his acts. He executes his own will with exact regularity. He does not vary. Hence, all his creatures may depend on regularity. It seems like law. The power in every case is the power of God. Law has no power. The law of gravitation has no power. Matter has no power. One of the primary lessons we learn in physics is the inertia of matter. Matter can not move, unless moved upon; nor stop of itself, when once in motion. Absolutely powerless! The power of attraction, which we may call a property of matter, is really the power of God. The effects are the results of power and intelligence. Law has neither power nor intelligence. Human law marks out the course man *should* pursue. Divine law records the course God *has* pursued. Human law must be enforced by all the executive power of the nation. God executes his own will, with perfect regularity; and, by courtesy of language, we call it "law." He is the great executor of the universe, not far removed, but proven present everywhere, by the power and wisdom necessary to produce the results. These results are found in the boundless universe, and in the microscopic world. They are found in the world far below the power of the most powerful microscope to detect. All the combinations of chemical elements are made, hidden from the eye of the microscope. Substances are dissolved and new combinations made, atoms are numbered, counted and combined with mathematical precision, and with an intelligence difficult for man to compute. No law could do this. Only a Being who has sufficient power and intelligence is equal to it. Law has no power, nor intelligence. Water is composed of two atoms of hydrogen and one of oxygen, combined with absolute precision everywhere. All chemical reactions require computations of an intelligent being. All nature teems with proofs that God is every where present. The elements in a high explosive are arranged instantly in new combinations, each atom taking its proper partners, in the proper proportion, with unerring precision. Countless calculations of the most difficult kind are made instantly and continually by the divine mind. Thus God's presence everywhere in the minutest forms of matter is clearly proved. It is a mathematical demonstration. God is not wearied by the care of worlds and suns, and systems and snow-drifts of stars on the highway of heaven, and takes just as perfect notice of atoms and electrons. They who think God is unable or unwilling to take care of the minutest division of matter as well as the rolling suns, must have a very diluted idea of God. It is now claimed that the atom, formerly believed to be the smallest division of matter, consists of 1740 parts. Sir Oliver Lodge says that the structure of an atom is as complex as that of a piano. This latest scientific discovery detects the power and wisdom of God, controlling, for ages, this minutest division of matter, undetected by the most powerful microscope.

It staggers one to think of the countless and difficult calculations that are made instantly by the divine mind in every part of the universe. The path of every snowflake that lazily pursues its tortuous course, and rests upon the lap of earth, is marked out, not by any law or agent, but by God himself. He calculates instantly the cyclone's path, the movement of every particle of air, the direction, velocity and path of every raindrop. A law could not do it.

The wisest man could not do it. But God can do it, with the ease with which the tempest carries a feather on its bosom, or the ocean floats a straw! Every second, about 16,000,000 tons of rain and snow fall to the earth; and God calculates the paths of the myriad flakes of snow and drops of rain instantly and unerringly.

The Conservation of Energy and the inter-convertibility of forces—light, heat, electricity,—taking place constantly everywhere, often on a stupendous scale, require bewildering calculations by an ever-present God. No energy, not even potential energy, can be lost in converting one force into another. It must be computed exactly.

Who but an infinite God could have calculated the enormous potential energy of the nebulous gases, required by contraction to cause the prodigious heat of a universe of suns?

The earth turns over noiselessly every 24 hours, carrying on its bosom, at the rate of 1000 mi. an hour, at dizzy heights, a most tenuous atmosphere, without a rustle, without the loss of a second in 1000 years. The earth with its satellite, is traveling around the sun at the rate of 18.5 mi. per second—75 times as fast as a cannon ball,—bearing a load of 6,000,000,000,000,000,000,000,000 tons, and arriving at a given point in its orbit, on exact time every tropical year. It has arrived so promptly on time following its elliptical course, at such a rate that the radius vector, a line from the sun to the earth, passes over equal areas in equal times, furnishing every moment an abtruse problem difficult for a scholar to solve. The orbit is so vast that it varies from a straight line, but 4 in. in 666 mi., the distance from Philadelphia to Chicago.

The sun also, with its family of worlds and their satellites, is plunging through space at the rate of 8.5 mi. per second; moreover, there are swarms of huge suns, many larger than ours, moving in straight-lines like a universe on a journey, and countless millions of suns in swiftest flight through the skies, whose orbits and rates of motion must all be calculated and controlled by a mind of amazing power and intelligence.

Is not the so-called "scientist" either a madman or a fool, who believes that all this can be accounted for, without the presence of a God of infinite power and intelligence?

Water contracts as the temperature falls. But when within four degrees of the freezing point, water expands and ice becomes lighter than water, and floats, and saves all bodies of water from becoming solid bodies of ice.

Who can say that God does not intervene, in this case, to save all life? It is a striking proof that God is not absent nor inactive.

Gravitation requires the computation of countless millions of the most complex and difficult problems, every instant, by the divine mind. The attraction of all matter for all other matter is in proportion directly to the mass and inversely to the square of the distance. The exact weight of every object is determined by the attraction of the earth and every particle thereof, the mountain that may be nearby, the elevation and altitude of the place, the attraction of the sun and the moon, and every star in heaven, even though too small to be computed by man, — all these are computed precisely by the divine mind. These innumerable calculations prove that God is everywhere. We are continually in the immediate awesome presence of an Infinite God.

Every computation that man ever made, was made long before by a great Intelligence, that excels all others combined. How intricate is the calculation of the divine mind, which causes the water of every ocean, sea, lake, pond, and vessel, when at rest, to correspond with the exact sphericity of the earth. In the face of innumerable and difficult calculations, — proofs of the intense activity of the divine mind, — who can be so reckless as to say that God is absent or inactive?

Not only does God make endless calculations in executing his will in the material universe, but in the intellectual, moral and spiritual world as well. We can not measure, with any human instruments, the amount of mental discipline and improvement, resulting from a certain amount of study. But God calculates unerringly the precise amount of mental discipline or improvement earned by every mental exertion. The amount is in precise proportion to the mental effort. The gain is definite, exact and unerring, the calculation is instantaneous, and beyond the power of the profoundest mathematician to compute. So also, the effect of every moral act, wish, desire, purpose, intention or affection, is instantly computed, and the moral character modified in exact proportion to their weight. If a man indulges in vice, he becomes vicious in proportion. If he commits a crime, he becomes more criminal in nature. Every theft is computed at its proper value. Every good and noble act ennobles the character in proportion to its worth. There is a settlement, every instant, and all deeds, wishes, desires, purposes, and affections go into the character, and affect it in precise proportion to their weight. Who but an infinite God, can keep all accounts of his innumerable creatures instantaneously, and have them complete, exact and unerring? No man, nor angel, nor "law," could do it. In like manner, every spiritual act, wish, purpose, motive, — all go in to make up the spiritual life of man, in exact proportion to their worth. Not all the mathematicians and scribes in the universe could together solve the problems, that the great intellect of the Supreme Ruler is solving every instant of time.

This theory of an absent or inactive God leaves no place for prayer, an almost universal instinct of mankind. If a blind, deaf, and dumb and helpless law is in control, it is useless to pray for help. All nations, races and peoples instinctively believe that God hears and answers prayer. This is a scientific fact with which evolutionists must reckon, even if it has a pious or otherwise offensive sound. No use to pray to an inexorable "law," which, like the gods of the heathen, can neither see, nor hear, nor taste, nor smell.

How unscientific then seems the following declaration of Darwin: "To my mind, it accords better with what we know of the laws impressed on matter [How could that be?] by the Creator, that the production and extinction of the past and present inhabitants of the world should have been due to secondary causes, like those determining the birth and death of the individual." It does not remove the First Great Cause from active control of the world to call his acts "secondary causes."

14
CHANCE OR DESIGN?

Evolution is the old heathen doctrine of chance. It professes to eliminate design and a personal active Creator. The theory of natural selection allows no design, no intelligence, no interference, no control, by the Creator. He does not interfere even by means of law. M. M. Metcalf, of Oberlin, O., (shades of Chas. G. Finney!), a prominent evolutionist, says, "The last stand was made by those who claim that supernatural agency intervenes in nature in such a way as to modify the natural order of events. When Darwin came to dislodge them from this, their last intrenchment, there was a fight." Yes! the fight will last while any one tries to substitute chance for the control of Almighty God.

The universe teems with countless evidences of intelligent design of the highest order, whether it is found in the starry heavens, or in the law and order of the atoms hiding from the most powerful microscope. All things came by chance or by design. They say there is no design. We wonder that the hand that wrote the lie was not palsied. It would be, if the same Creator that filled every muscle, nerve, bone, and tissue of the sacrilegious hand, with numberless proofs of design, were not a long-suffering and merciful God.

Prof. Vernon Kellogg says: "Darwinism may be defined as a certain rational causo-mechanical (hence non-teleologic) explanation of the origin of species." Translated into plain English, this euphemistic expression means that Darwinism excludes all design and control by a Creator. Chance pure and simple. All species originated by chance, without interference by a supreme Being. This senseless doctrine of chance has been condemned by man in every age.

We can only note a few of the evidences of design, found in bewildering numbers in every part of God's great creation.

The Human Body. Can evolutionists imagine how the human body could be crammed fuller of the clearest proofs of the most intelligent design, indicating a mind of the highest order? Many of the most remarkable inventions of man were suggested by the wonderful contrivances found in the human body. Yet they say this marvelous piece of ingenuity did not come from the hand of the Creator but was developed by blind chance or "natural laws," without a trace of intelligent design by the Creator, or by man or beast. The human body can no more be a product of chance or causo-mechanical evolution than a Hoe printing press, or Milton's Paradise Lost.

On high medical authority, we are told that there are in the human body 600 muscles, 1000 miles of blood vessels, and 550 arteries important enough to name. The skin, spread out, would cover 16 square feet. It has 1,500,000 sweat glands which spread out on one surface, would occupy over 10,000 sq. ft., and would cover 5 city lots, 20 x 100 ft. The lungs are composed of 700,000,000 cells of honey comb, all of which we use in breathing,—equal to a flat surface of 2,000 square feet, which would cover a city lot. In 70 years, the heart beats 2,500,000,000 times, and lifts 500,000 tons of blood. The nervous system, controlled by the brain has 3,000,000,000,000 nerve cells, 9,200,000,000 of which are in the cortex or covering of the brain alone. In the blood are 30,000,000 white corpuscles, and 180,000,000,000,000 red ones. Almost 3 pints of saliva are swallowed every day, and the stomach generates daily from 5 to 10 quarts of gastric juice, which digests food and destroys germs. Two gallons daily! It is easy also to believe that the "very hairs of our heads are numbered,"—about 250,000.

Yet many an upstart, with thousands of the most marvelous contrivances in his own body, is ready to shout that there is no God and no design, or that there has been no interference since creation, and that our bodies have reached the dizzy heights of perfection, without intelligence, purpose or design. Absurd in the highest degree! "We are fearfully and wonderfully made."

The Eye. Darwin says, "To suppose that the eye with all its inimitable contrivances for adjusting the focus to different distances, for admitting different amounts of light, and for the correction of spherical and chromatic aberration, could have been formed by natural selection, *seems, I frankly confess absurd in the highest degree.*" (Italics ours). After admitting that it "seems absurd in the highest degree," he proceeds, as if it were certainly true. Darwin has been admired for his candor, but not for his consistency. After admitting that an objection is insuperable, he goes on as if it had little or no weight. And many of his followers take the same unscientific attitude. They try to establish their theory in spite of overwhelming arguments.

"Reason tells me," he says, "that if numerous gradations from a simple and imperfect eye, to one complex and perfect, can be shown to exist, such gradation being useful to its possessor, as is certainly the case" (certainly?), "if further," he continues, "the eye varies and the variations be inherited, as is likewise certainly the case" (most modern evolutionists say certainly *not* the case; what, if variations are unfavorable?); "And if such variations should be useful, (what if not useful?) to any animal under changing conditions of life, then the difficulty of believing that a perfect and complex eye *could* be formed by natural selection, *though insuperable to the imagination* (Italics ours) should not be considered as subversive of the theory"!! Darwin undertakes a task far too great for his mighty genius. "Believing that a perfect and complex eye could be formed" is many moral leagues from proving that it was so formed. We must have stronger proof than sufficient to lead us to believe that such an eye could possibly be so formed. All proof is exhausted in the struggle to prove the possibility of the formation of so marvelous an eye, to say nothing of the probability, much less the certainty required by science. We hold evolutionists to the necessity of proving that the eye was *certainly* so formed. We demand it. Otherwise, we shall certainly "consider it subversive of the theory." And if acquired by one species, how could it benefit another species? But we must contest the claim that the wonderful eye of man and animals *could* have been formed by evolution. Darwin's whole theory aims to account for all creation, with its super-abundant evidences of design, by natural selection, which works without design and without intelligence. The theory is founded upon the monstrous assumption that unintelligent animals and plants, can, by aimless effort arrive at such perfection as the organs of the human body, exceeding anything in mechanical contrivance, invented to date by the genius of man. Indeed, that wonderful invention of the telescope is but a poor imitation of the eye, and does not begin to equal it in marvelous design. Who would say that the telescope might have been constructed by chance, or the fortuitous concurrence of atoms, or by natural selection, or any other attempted method of blotting out the great intelligent Designer of the universe? It not only "*seems* absurd in the highest degree," but certainly *is*, and is fatal to the theory.

The eye is so wonderful in its powers, and delicate adjustments, that we stand amazed at the evidences of design, and at the wisdom of the Maker of the eye, far exceeding the highest inventive genius of man. To say that this is the result of "natural selection," is absurd and ridiculous. Evolution eliminates design, mind, and an active and ever present God, and substitutes blind chance or natural selection, dubs it "science" and asks the world to believe it!

According to the evolution theory, the gain in the mechanism of the eye causes its possessors to survive, and others to die. Is that true? Are there not many species that survive, whose eyes are less perfect than the eye of man? Indeed, it is claimed that many animals have eyes superior to man. If so, why did man survive and become the dominant species, with eyes less perfect? The compound eyes of some species are superior in some respects, as every one knows, who has ever tried to slip up on a fly. A scientist says that fleas have such perfect vision that the darkness under the bed clothes is to them a glaring light.

Darwin makes a fatal admission, when he says, "To arrive, however, at a conclusion regarding the formation of the eye with all its marvelous yet not absolutely perfect characters, it is indispensable that the reason should conquer the imagination; But I *have felt the difficulty far too keenly to be surprised at others hesitating to extend the principle of natural selection to so startling a length.*" (Italics ours). No wonder the reason and judgment of mankind revolts against such a theory and that so many evolutionists themselves reject it.

Three or four per cent. of the population are color blind—"red-blind" —and are not able to distinguish the color of the green leaves from that of the red ripe cherries. Can it be possible that the eye becomes more perfect, because those who had less perfect eyes perished, and only those who could recognize colors survive until color blindness is finally eliminated? Is such a doctrine scientific? Is it more reasonable to believe it than to believe that an infinitely wise and powerful God created this organ of marvelous value and beauty? Of course, the ability to recognize color is only one of the many perfections of the eye.

Evolution is made so much more incredible, because it teaches that every permanent improvement in the eye is made at the expense of multitudes of individuals that perished because of the lack of the improvement. The defect perished only because all individuals afflicted with it perished. Is this true?

The bureau of education of the U.S. government reports that, of *22,000,000* school children examined, 5,000,000 have defective eyes; 1,000,000, defective hearing; 1,000,000 have active tuberculosis; 250,000, heart trouble; 3,000,000 to 5,000,000 are underfed; total, 12,250,000,—more than half. Must all these defectives perish in order that man may reach perfection? Less than half are the "fittest" and they only could survive.

Location of organs. But if the evolutionist *could* convince the thoughtful student that the marvelous eye could have been so formed, by blind chance or natural selection, how could he account for the advantageous location of the eye and other organs? While we can not well name a fraction small enough to express the mathematical probability of the formation of the eye, the ear, and other organs of the body, we easily can compute the fraction of the probability of their location, though very small. In the passage quoted from Darwin, he begins with the simple eye, but does not say how the eye originated. Hon. William J. Bryan in his book, "In His Image," p. 97, says, "But how does the evolutionist explain the eye, when he leaves God out? Here is the only guess that I have seen,—if you find any others, I shall be glad to know of them, as I am collecting the guesses of the evolutionists. The evolutionist guesses that there was a time when eyes were unknown—that is a necessary part of the hypothesis. And since the eye is a universal possession, among living things, the evolutionist guesses that it came into being,—not by design or act of God—I will give you the guess,—a piece of pigment, or as some say, a freckle, appeared upon the skin of an animal that had no eyes. This piece of pigment or freckle converged the rays of the sun upon that spot, and when the little animal felt the heat on that spot, it turned the spot to the sun to get more heat. This increased heat irritated the skin,—so the evolutionists guess—and a nerve came there and out of the nerve came the eye. Can you beat it? But this only accounts for one eye; there must have been another piece of pigment or freckle soon afterward, and just in the right place in order to give the animal two eyes."

Now assuming, what seems an utter impossibility, that the wonderful mechanism of the eye can be accounted for by chance or natural selection (another name for chance since design is excluded), how can we account for the *location* of the eyes, and, in fact, of all the other organs of the body? We can easily calculate the mathematical probability on the basis of natural selection. There are from 2500 to 3500 square inches of surface to the human body, a space easily 3000 times the space occupied by an eye. The eye, by the laws of probability, is just as likely to be located any where else, and has one chance out of 3000 to be located where it is. But out, of our abundant margin, we will concede the chance to be one out of 1000, and hence its mathematical probability is .001. For mathematical probability includes possibility and even improbability. The compound probability of two things happening together is ascertained by multiplying together their fractions of probability. Now the probability of the location of the second eye where it is, also is .001. And the compound probability of the location of both eyes where they are, is .001 x .001 or .000,001. In like manner, the probability of the location of each ear where it is, is .001, and of the two ears .000,001. The compound probability of the location of two eyes and two ears where they are, is .000001 x .000001 or .000,000,000,001. The two eyes and two ears have but one chance out of a trillion or a million million to be located where they are. The location of the mouth, the nose, and every organ of the body diminishes this probability a thousand fold. We are speaking mildly when we say that this calculation proves that the evolution of the body, by chance or natural selection, has not one chance in a million to be true. So ruthlessly does the pure and reliable science of mathematics shatter the theory of evolution, which so called scientists claim is as firmly established as the law of gravitation.

Concerning the wild guess of the development of the legs, we again quote from Mr. Bryan, "In His Image," p. 98: "And according to the evolutionist, there was a time when animals had no legs, and so the legs came by accident. How? Well, the guess is that a little animal was wiggling along on its belly one day, when it discovered a wart—it just happened so,—and it was in the right place to be used to aid it in locomotion; so, it came to depend upon the wart, and use finally developed it into a leg. And then another wart, and another leg, at the proper time—by accident—and accidentally in the proper place. Is it not astonishing that any person, intelligent enough to teach school, would talk such tommyrot to students, and look serious while doing so?"

Some one has counted that Darwin has used phrases of doubt, like "We may well suppose," 800 times in his two principal works. The whole theory is built up on guesses and suppositions. "Let us suppose" that each guess is 95 per cent certain, which is far higher than the average or any. The compound probability would equal .95 raised to the 800th power which would be .000,000,000,000,000,006,281 which means there are 6 chances out of a quintillion that evolution is true. Since not all of these 800 suppositions are dependent upon each other, we are willing to multiply this result by 10,000,000,000 which still shows that the theory has less than one chance in a million to be true. Darwin himself says, "The belief that an organ so perfect as the eye could have been formed by natural selection, is more than enough to STAGGER ANY ONE." Yet he and his followers refuse to be "staggered," and proceed to argue as if this unanswerable objection had little or no weight. *Any hypothesis is weakened or damaged by every support that is an uncertain guess.* Gravitation has no such support.

Mr. Alfred W. McCann, in his great volume "God or Gorilla," shows that H. G. Wells, the novelist *alias* historian(?), in his "Outline of History," uses 103 pages to show man's descent from an ape-like ancestry, and employs 96 expressions of doubt or uncertainty, such as "probably," "perhaps," "possibly," etc. He does not hesitate to endorse the wildest guesses of the evolutionists, and sits upon the top of this pyramid of doubt, and proclaims, *ex cathedra*, apparently without a blush, of our ancestors: "It was half-ape, half-monkey [elsewhere, he says the lemur was our ancestor]. It clambered about the trees and ran, and probably ran well, on its hind legs upon the ground. It was small brained by our present standards, but it had clever hands with which it handled fruit and beat nuts upon the rocks, and perhaps caught up sticks and stones to smite its fellows. IT WAS OUR ANCESTOR."!!!

And he does not hesitate to give a picture of our ancestor drawn by an artist 500,000 years after its death. Yet this book so dangerous, so anti-christian, and so untruthful concerning the origin of man, is recommended by careless librarians, by scholars, and even by Christians. It will take a long time to erase from the mind of the youth, the false teachings of this book. It is one of the most cunningly devised plans ever attempted to teach infidelity and atheism in the name of history.

Plans for man prove design. All nature is crowded with evidence that God intended to create man. He made great preparation for his coming. He provided many things useful to man but to no other species. Veins of coal, almost innumerable—the canned sunshine of past ages—, are placed near the earth's surface, accessible for man, when needed for his use. Of no value whatever to any other species, because they can not make or replenish a fire. A colored preacher did not miss the mark, when he said, "God stored his coal in his great big cellar for the use of man." The man who fills his own cellar with provisions for the winter exhibits no more foresight or design.

The oil and gas were also evidently stored away in the earth for the use of man. It is worth nothing to animals. Over 41,000,000,000 gallons of oil were consumed in the U.S. in 1924.

All the other minerals likewise were stored in the earth for the use of man alone,—iron, copper, gold, silver, all the valuable minerals,—knowing that man would make use of them. The most precious and most useful minerals are of no value whatever to any species of animals. God foresaw the marvelous inventions of the present and the future, and provided the means ages ahead of time. The universe is crowded so full of design, that there is no room for chance or natural selection.

15
EVOLUTION ATHEISTIC

Evolution harmonizes with atheism and kindred false theories. This raises a presumption against its truth, as falsehood does not agree with the truth. It is reconcilable with infidelity and atheism, but not with Christianity. Many, like Prof. Coulter, of the Chicago University, endeavor to show that evolution is reconcilable with *religion*—and he does show that it harmonizes with the religion of deism or infidelity. No one doubts that evolution harmonizes with atheism or the religion of Thomas Paine. But why should we be anxious to reconcile it with Christianity, when there is so little truth to support it?

Many evolutionists are atheists. Some believe in the eternity of matter. This can not be. Both mind and matter can not be eternal. Mind controls matter; and not matter, mind. Hence the mind of God created matter.

Some believe the universe came into being by its own power, though that can not be. Power or force cannot create itself. It must be attached directly or indirectly to a person. No force can be disconnected from its cause. Detached force is unthinkable. All force in the universe can be traced to God. Much of the physical power of the earth can be traced to the sun,—storms, cataracts, steam, electricity,—and the sun gets its power from God. Gravitation, extensive as the universe, is but the power of God in each case.

The total force in the universe *is* beyond calculation. It is a part of the power of Almighty God. It approaches infinity. All heat is convertible into power, and power into heat. Heat, when converted into power, moves the mighty engines. The power of Niagara may be converted into heat and light. The sun had lifted the waters of the whole Niagara River, and the lakes far above the Falls. Its power is enormous. It lifts up over 1,000,000,000,000 tons of water to the clouds every day,—more than all the rivers and streams pour into the seas. The sun equals in size a pile of more than a million worlds like ours. Every square yard of surface of this enormous sphere, has enough heat to push a great liner across the sea,—as much power as in many tons of coal. The amount of heat in the surface of the sun, consisting of more than 2,284,000,000,000 sq. mi., can hardly be imagined. The heat of one sq. mi.

(3,097,600 sq. yds.) would drive 3,000,000 ships across the sea,—150 times as many as are afloat. More than 2,200,000,000 times as much heat as the earth receives, goes out into space. And this enormous amount of heat is but a poor fraction of the heat of 400,000,000 suns, few of which are so small as ours.

A single star, Betelguese, has recently been computed to be 215,000,000 mi. in diameter, and therefore larger than 10,000,000 suns like ours. A still more recent computation shows stars even larger. Antares is 390,000,000 mi. in diameter, equal, to 91,125,000 suns, or 136,687,500,000,000 worlds. If our sun were in the centre of this sun, it would extend beyond the orbit of Mars. Alpha Hercules is 300,000,000 mi. in diameter. Some stars are so far away that it takes light 60,000 years to reach us, at the rate of 186,000 mi. in a second. Some say there are 400,000,000 enormous suns. Compute, if you can, the sum total of the power causing the light and heat, and the power of gravitation controlling these vast swarms of stars. All this power is the power of God, and a weak fraction of the total. This power could not originate itself. It could not grow. It could not come by evolution. It could not come by chance.

The doctrine of the Conservation of Force, accepted by scientists, proves that no part of force can be lost. A God of infinite power is required to create, maintain and control this vast universe. Force can no more create itself than matter. God must create and preserve both. It takes almighty power to maintain the universe in existence, as well as to create it.

If atheism be true, then, if there was even one germ to start with, as most admit, it must have created itself, unless the absurd claim that it came from another world, riding on a meteorite, be entertained. If such a foolish assumption were possible, it would require a God to create it in another world.

"The fool hath said in his heart, 'No God'." Some translators would supply the words omitted by the Hebrew, and make it read: "The fool hath said in his heart, '*There is* no God'." Others, "The fool hath said in his heart, '*I wish there were* no God'." It is hard to tell which is the bigger fool, the man who refuses to see the countless evidences of design, proving His existence; or the man who refuses to see the terrible wreck of the great universe, and the awful chaos that would result if there were no God. We can imagine only one greater fool than either: The man who thinks he can get the world to believe, under cover of evolution, that there is no God, and that all things were evolved by chance, even though it be camouflaged by the terms "natural selection" or "natural law."

Atheism implies spontaneous generation, which is entirely without proof. Indeed, if spontaneous generation were possible at the beginning of life, it is possible now, and has been possible during all the ages. But no proof of it has been given. On the contrary, all efforts to secure, by chemistry, the lowest forms of life from dead matter have been without avail. Dr. Leib, of Chicago University, made earnest efforts to do so. He failed utterly. If nature, aided by the genius of man, can not now produce the lowest forms of life from matter, how could it ever have been done? Prof. Huxley filled jars with sterilized water, and placed in it sterilized vegetation, and sealed them up, and after 30 years, no life was seen, disproving spontaneous generation. Pasteur proved that, if milk were sterilized, there would be no development of life by spontaneous generation. This discovery was of immense practical value, making milk safe to use. Prof. Tyndall, the distinguished physicist, said: "If matter is what the world believes it to be, materialism, spontaneous generation, and evolution, or development, are absurdities too monstrous to be entertained by any sane mind." Dr. Clark Maxwell, another distinguished physicist, says, "I have examined all [theories of evolution] and have found that every one must have a God to make it work." *L'Univers* says: "When hypotheses tend to nothing less than the shutting out of God from the thoughts and hearts of men, and the diffusion of the leprosy of materialism, the savant who invents and propagates them is either a criminal or a fool." Even Darwin seems to be conscious of a designing mind when he says, "It is difficult to avoid personifying the word Nature. But I mean by nature only the aggregate action and product of many natural laws." A futile effort to exclude God. Who made these laws?

Can a theory that is consistent with false theories, like chance and atheism be true? Truth is consistent with truth, but not with falsehood. We can judge a theory by the company it keeps. Evolution naturally affiliates with false theories rather than with the truth. It favors infidelity and atheism. A theory in perfect harmony with manifest error, raises a presumption against its truth. Evolution seems to have a natural attraction for erroneous hypotheses and manifests the closest kinship with impossible theories. This is not a mark of a true theory.

So baneful has been the effect of teaching evolution as a proven hypothesis, that multitudes have been led into infidelity and atheism. Prof. James H. Leuba, of Bryn Mawr College, Pa., sent a questionaire to 1000 of the most prominent scientists teaching sciences relating to evolution. The replies indicate that more than one-half do not believe in a personal God, nor the immortality of the soul,—beliefs almost universal even in the heathen world. So pernicious is this doctrine of evolution that more than one-half of the professors who teach it and kindred subjects, are infidels and atheists and farther from God than the ignorant heathen. And while we are happy in the conviction that the great majority of professors and teachers of other subjects are Christians, yet one or two atheists or infidels are sufficient to make havoc of the faith of many, in a great college or university.

A doctrine so abhorrent to the conscience, so contrary to the well nigh universal belief, and so fruitful of evil, certainly can not be true. Small wonder is it that students are fast becoming infidels and atheists, and we shudder as we think of the coming generation. A great responsibility rests upon the authorities who employ such teachers.

The answers of the students in seven large representative colleges and universities to Prof. Leuba's questionaire, show that while only 15% of the Freshmen have abandoned the Christian religion, 30% of the Juniors and over 40% of the Seniors have abandoned the Christian faith. Note the steady and rapid growth of infidelity and atheism as a result of this pernicious theory.

Will Christian parents patronize or support or endow institutions that give an education that is worse than worthless? What the colleges teach today the world will believe tomorrow.

Atheism, under its own name, has never had many to embrace it. Its only hope is to be tolerated and believed under some other name. In Russia, no man is allowed to belong to the ruling (Communist) party unless he is an atheist. It will be a sorry world when "scientific" atheism wins, under the name of evolution.

No one has a moral right to believe what is false, much less to teach it, under the specious plea of freedom of thought.

It is the privilege and duty of parents to send their children to institutions that are safe.

Nathan Leopold, Jr., and Richard Loeb kidnapped and cruelly murdered Robert Franks. Both were brilliant scholars and atheists. Both graduates of universities, though minors, and both were taking a post-graduate course in the University of Chicago. It is asserted and widely believed that they were encouraged in their atheistic belief by the teaching of evolution and modernism, and were thus prepared to commit a crime that shocked the world.

Most of the writers who advocated evolution became atheists or infidels; most of the professors who teach it, believe neither in God nor the immortality of the soul; and the number of students discarding Christianity rose from 15% in the Freshman year to 40% in the Senior. What more proof is needed?

16
BRUTE DESCENT IMPOSSIBLE

According to Prof. R. S. Lull and other evolutionists, "The skull of the pithecanthropus is characterized by a limited capacity of about two-thirds that of a man." Assuming that this skull is that of a normal creature of that age, as is done in all the arguments of "our friends, the enemy," then the pithecanthropus must have lived 20,000,000 years ago, one-third the period assigned to life. They claim the pithecanthropus lived 750,000 years ago; later the guess is reduced to 375,000. Does any one in his senses believe that an ape-human animal developed one-third of the normal human brain in 375,000 or 750,000 years, when it took 59,250,000 years to develop two-thirds of the brain? If one-third of the normal brain developed in the last 750,000 years, the rate of development must have been 39.5 times as great as in the preceding 59,250,000 years. If one-third developed in the last 375,000 years, the rate of development must have been 78 times as rapid as in the preceding 59,625,000 years. This is incredible. If life began 500,000,000 years ago, and one-third the brain developed in the last 750,000 years, the rate must have been 332 times as rapid as in the preceding 499,250,000 years; and 666 times as rapid in 375,000 years as in the preceding 499,625,000 years. All these guesses are clearly impossible.

But the agile evolutionist may try to escape the death sentence of mathematics and the condemnation of reason, by saying that the brain developed more rapidly than the rest of the body. But he is estopped from that claim, by the statement of this same Prof. R. S. Lull: "The brain, especially the type of brain found in the higher human races, must have been *very* slow of development." If so, the pithecanthropus must have lived more than 20,000,000 years ago! So swiftly does inexorable mathematics upset this reckless theory.

This calculation has been made upon the basis of the estimate of 60,000,000 years since life began, taken from Prof. H. H. Newman in "Readings in Evolution," p. 68. But, seeing that even this great estimate of the period of life is not sufficient for evolution, in a private letter to the writer, Prof. Newman raises his guess to 500,000,000 years. In that case, the pithecanthropus must have lived one-third of 500,000,000, or 166,666,666 years ago. And, if we are reckless enough to admit the "moderate estimate" of 1,000,000,000 years, gravely suggested by Prof. Russell, of Princeton University, it must have lived 333,333,333 years ago. These reckless estimates seem removed, by the whole diameter of reason, from even a respectable guess. Every new guess seems to make their case more hopeless. And any guess that they can make, out of harmony with the Scripture statement, can be disproved by cold mathematics. In like manner, if the Piltdown man had the estimated brain capacity of 1070 c.c., instead of the normal 1500 c.c., this fabricated creature must have lived about 17,200,000 years ago, if life began 60,000,000 years ago; and 143,333,333 years ago, if life began 500,000,000 years ago; (c.c. = cubic centimeters).

Prof. Schaaffhausen, the discoverer, estimated the capacity of the Neanderthal man at 1033 c.c. Then he must have lived 18,680,000 years ago, if we accept the 60,000,000 year period; and 311,333,333 years ago, if we accept Prof. Russell's guess of 1,000,000,000 years.

And in all these long ages, fragments of only four skeletons of very doubtful character have been found, and upon this flimsy proof, the youth of our land are expected by self-styled "scientists" to believe it, even though it leads them into infidelity and atheism, and causes the loss of their souls.

Let us take another view. Let us assume that the pithecanthropus really lived 750,000 years ago, as claimed, which is 1.25% of 60,000,000 years. Therefore, its brain capacity then should have been 98.75% normal, or 1481.25 c.c. or 18.75 c.c. less than the normal 1500 c.c. Also 750,000 years is only .15% of 500,000,000 years; hence in that case, the brain should have been 99.85% normal, or 1497.75 c.c. In either case, the intelligence must have excelled that of many nations and races. All these calculations prove positively that no such creatures as these four alleged ape-men ever could have lived in the age assigned to them; or, if so, that none could have had, at that time, the low brain capacity claimed. Q. E. D.

Is it not plain that for the last 2,000,000 years out of 60,000,000 years, the developing human race must have been over 29/30 or 96 2/3% normal, in intelligence, morality, and spirituality? This is greater than that of many peoples today. With this high degree of intelligence, man was capable of great inventions and discoveries. Not a single monument remains. We would expect some great monument like the pyramids of Egypt. A race with such advancement, for so many years would have been able to reach the heights of invention, discovery, and learning of the present age. Not a whit of evidence comes down to us.

If 2,000,000 years ago, man had the same skull capacity as the ape, 600 c.c., he has gained 900 c.c. in 2,000,000 years, and only 600 c.c. in 58,000,000 years. His improvement in the last 2,000,000 years, must have been 43.5 times as rapid as during the preceding 58,000,000 years; or 373.5 times as rapid as during the preceding 498,000,000 years. How was that possible?

17
EIGHT IMPASSABLE GULFS

The evolution theory, stretching from matter to man, is impossible, because of many impassable gulfs. Some of these impassable gulfs are:—

1. Between the living and non-living or dead matter;

2. Between the vegetable and the animal kingdoms;

3. Between the invertebrates and the vertebrates;

4. Between marine animals and amphibians;

5. Between amphibians and reptiles;

6. Between reptiles and birds;

7. Between reptiles and mammals;

8. Between mammals and the human body;

9. Between soulless simians and the soul of man, bearing the image of God.

There is not a scrap of evidence that these gulfs have ever been crossed. In the scheme, the material must become living by spontaneous generation; some plants must become invertebrate animals; some invertebrates must become vertebrates; some marine animals must become amphibians; some amphibians must become reptiles; some reptiles must become mammals; some mammals must become humans; some senseless, soulless simians must acquire a soul and become spiritual enough to bear the image of God.

There is no convincing proof that any of these great and incredible advances were ever made. If we estimate the probability of each transmutation at 10%, which is too high, then the probability that all these changes up to man were made is .1 raised to the 8th power, .00000001. Therefore, there is not more than one chance out of 100,000,000 that these 8 changes were made. And if we estimate the probability of each great change at .001, which is doubtless still too high, the probability that man took these 8 great steps of evolution is one out of 1,000,000,000,000,000,000,000,

000,000, or a million, million, million, million. If we estimate the probability of each change even at 60%, which is far above all reason, the probability of man's evolution through these 8 changes is only 1 out of 60, which marks an improbability close to an impossibility. The highest estimate we can reasonably make, destroys all hope that man or even any other species could have come by evolution. Few persons realize how improbable an event is made which depends upon a number of possibilities or even probabilities, until calculated by the rule of Compound Mathematical Probability.

Imagine the Copernican or the gravitation theory depending on a number of possibilities or probabilities! No true theory is built on such an uncertain foundation.

But, if the evolutionists could prove that 7 out of 8 of the great changes certainly did occur, but failed to prove the 8th, they would lose their case. But they have failed in all. They must prove all to win. There is not the slightest probability that any one of these changes ever occurred. Hence, the evolution of man from this long line of alleged ancestors is an absolute impossibility. Q. E. D.

None of these changes is *now* occurring. There is no spontaneous generation now. Darwin himself said that spontaneous generation in the past was "absolutely inconceivable." No reptiles are becoming mammals, none becoming birds, no apes or monkeys are becoming men. No species is now transmuted into another, no new species arises. Is not this proof enough that such great changes never occurred?

Moreover, if dead matter caused one living germ, why did it not cause more? If some reptiles developed into mammals, and birds, why not all? If one family of simians became human, why not others? Why not at least become anthropoids? Why did all other members of the simian family not become at least part human? Why have they remained stationary?

Besides, we have with us yet the invertebrates that have not yet become vertebrates; marine animals that have not become amphibians; amphibians that have not become reptiles; reptiles that have become neither mammals nor birds, and a multitude of simians that have not become human, and are not moving toward man either in bodily form or intelligence or spirituality. We have the one-celled amoeba, the microscopic animals, and the lowest forms of animal life. If the great law of progress and advancement to higher forms has prevailed for so many million years, there should be none but the highest species. All should have reached the status of human beings and there should be none of the lower forms of life which are so abundant. Changes so radical and vast, stretching through so many ages, would require millions of connecting links. If reptiles became hairy mammals, we would expect fossils of thousands, if not millions, in the transition state. If some reptiles were changed into the 12,000 species of birds, we would expect countless fossils, part reptile, part bird. Only one is claimed, the archæopteryx (ancient bird), two specimens of which are known, which had a feathered tail, and which is only a slight modification of other birds. Many other birds have departed farther from the normal. There should be millions of fossils in the transition state if the theory were true. We have proven elsewhere that there is no credible evidence of links connecting man with the monkey family. There would have been many millions. We have shown, at length, that some of these great changes, especially the Evolution of man from the brute, could never have occurred. No one of these nine great advances was ever made, but it will suffice to examine now, as examples, two alleged great changes, reptiles into mammals, and reptiles into birds.

1. Evolutionists say that mammals are descended from some reptiles, unknown, of course, and birds from others, also unknown. Mammals differ from reptiles in having breasts (Latin, mammae), a four chambered heart instead of three, a coat of hair or fur or wool, and a womb for the young. The temperature of the blood of reptiles is as low as 60 and even 40 degrees, since the temperature of the blood is about the same as the environment, sometimes approaching the freezing point. But mammals have a temperature approaching 100°. We are to believe that one progressive branch of reptiles, which passed through the sieve of natural selection, during the Permian Ice Age, was capable of being adapted to the colder climate. But this mighty chasm between reptiles and mammals was crossed unaided by any external interference, unaided by God; then the mammals groped their way, without intelligence or design, up to man! The difficulties are too great to satisfy the serious student. No satisfactory explanation has been given. No fossils, part reptile, part mammal, have been found. We would naturally expect millions of them. Evidently none ever existed. How could such radical changes be brought about? What caused the development of hair, fur and wool? The change in the heart, and the temperature, the formation of the mammae and of the womb? There is no evidence of such change. But it is necessary to the scheme.

2. Some reptiles became birds, they say; whether a pair for each of the 12,000 species of birds or one pair for all, we can not learn. For nobody knows. They would like for us to believe that these cold-blooded reptiles with a temperature of 40 to 60 degrees became birds with a temperature as high as 107; that wings and feathers were developed, which must have been perfectly useless through the long ages during which they were developing; that the wonderful contrivances in the wings and feathers were made by senseless reptiles that did not know what they were doing. Reptiles have

a three-chambered heart, making them cold-blooded. Birds have a four-chambered heart, and a temperature higher than that of man. Reptiles left their eggs to hatch in the sun. Birds, by a fine instinct, built their nests with care. Some reptiles have 4 feet, some 2, some none. All birds have two feet. The bird's structure is so well suited for flight and shows the marks of design so clearly, that the clumsy aeroplane is but a poor imitation. Yet to link the 12,000 species of birds to their unknown reptilian ancestors, they show us two fossils of the archæopteryx, as the sum total of the evidence showing the transition from reptiles to birds. The fossil varies slightly but not essentially from other birds. It has a feathered tail, some teeth and claws. It is probably not a connecting link at all, and if it were, we would expect a million fossils of connecting links. All these nine transmutations are devoid of a single sure connecting link, when we would expect millions in every case. These facts prove that evolution is a delusion and an absurdity.

18
ANCESTRAL APES AND MONKEYS

Many have taught that man was descended from an ape or monkey. Evolutionists, ashamed of a doctrine so repugnant to all reason and so revolting to mankind, vainly imagine they can escape the odium of such a view, by declaring that man is not descended from an ape or monkey, but that all the primates including all monkeys, apes, and man, sprang from a common ancestor. Of this alleged ancestor *not a single fossil remains*. Dr. Chapin, Social Evolution, page 39, says: "When the doctrine of the descent of man was first advanced, superficial and popular writers immediately jumped at the conclusion that naturalists believed that man was descended from the monkey. This, of course, is quite absurd, as man obviously could not be descended from a form of life now living. The ape and the monkey family, together with man are probably (?) descended from some generalized ape-like form long since perished from the earth." Suppose this absurd and unsupported guess to be correct. Then the gorillas, chimpanzees, gibbons, orang-outangs and other apes; the baboons and other monkeys; and the lemurs and man were brothers and sisters, or otherwise closely related, and all were descended immediately or nearly so from a common ancestor *lower than any*. Where is the comfort or gain? Moreover, all the members of this primate family must have inter-bred for ages, until, according to the theory, they became distinct species. Therefore, the ancestors of man, for ages, must have been descended from all these members of the primate family, and are thus the offspring of *all* these repulsive brutes, and the blood of them all is in our veins! In attempting to rescue us from the ape as our ancestor, they have shown that we are descendants of the whole monkey family and every species of ape and of many of their more disreputable relatives also. Great is evolution!

It certainly would be impossible for one single pair to have become the ancestors of the human race, without mixing and interbreeding with their kindred primates. Where are the descendants of these mongrel breeds, part monkey and part man? We would expect all gradations of mixed animals from monkey to man. "Two or three millions of years ago an enormous family of monkeys spread over Europe, Asia and Africa." All related, many our ancestors.

Why did not some other species of the primates equal or excel man or advance part way between man and the brute? Why are they not now becoming human? It is plain to the sincere student that the evolution of man from the brute is only the product of the imagination of those who wish to deny special creation and exclude God from his universe.

The slight external resemblance between man and the ape family is more than offset by structural differences which deny kinship. Alfred McCann in his great book "God—or Gorilla" says, p. 24, "Man has 12 pairs of ribs; the gibbon and chimpanzee, 13; man has 12 dorsal vertebrae; the chimpanzee and gorilla, 13; the gibbon, 14. The gorilla has massive spines on the cervical vertebrae above the scapula"; and, like the other quadrumana (4-handed animals) has an opposable thumb on the hind foot. There are wide differences in the shape of the skull, thorax, femur, and even the liver. The skeleton of the brutes is much more massive. On the tips of the fingers and thumbs of the human hand are lines arranged in whorls, for identification. In monkeys, the lines are parallel on the finger tips, but whorls on the palm. Is it possible that man and such brutes came from the same parents?

19
A STAGGERING SPECULATION

The theory that all plants and animals have descended from one primordial germ, is staggering to the mind. If so, how was it? Did this original germ split in two, like some disease germs, one of them the beginning of plant life, and the other the head of all animal life? Or, did vegetation only, grow from this first germ for ages, and then some of it turn into species of animals? As if the guess were worthy of attention, some are ready to assert that early vegetation Algae turned into animals. Did plants become animals somewhere along the way? Or did animals, somewhere along the way, turn into plants? How long did they interbreed before the gap became too wide? Where are the descendants of the union between plants and animals? If animals were first developed from this first germ, what did they live on while there was no vegetation? What folly is like the folly of the evolutionist who claims that such weird speculation is science?

Great gaps between the principal divisions of the animal world are fatal to this speculation, which rests upon nothing but the wish that it were so. Links are lacking between marine and amphibian animals; reptiles and birds; reptiles and mammals; between apes and man. Of course, we would find fossils of millions of these links if there were any. The missing links are necessary to the scheme. Is there one chance in a million that evolution is a true hypothesis?

20
SEX

Can the evolutionist explain the origin of sex? Starting with one germ or even a few germs, reproduction must have been by division for a time. If the germ that became the head of all plant life, reproduced by division, when did it begin to reproduce by seeds?

It is still more difficult to explain when sex life began in animals. There could have been no sex life at first, and perhaps for ages. They can not tell us when the animals, by chance, acquired the wonderful adaptation of the sexual life. They have no evidence whatever. Their guess is no better than that of others. It passes credulity to believe that the sexual life, with all its marvelous design, was reached by the invention of irrational animals, when man, with all his powers of reason, invention, and discovery, is helpless even to understand the great wisdom and power that brought it about.

Can blind chance, or aimless effort by senseless brutes, accomplish more than the amazing design of an infinitely wise and powerful God?

How was the progeny of mammals kept alive, during the ages required for the slow development of the mammae?

21
MAN HAIRLESS AND TAILLESS

How did man become a hairless animal? is a hard question for evolutionists. Any scientific theory must be ready to give an account of all phenomena. A hypothesis to explain the origin of man must explain all the facts. How did man become a hairless animal? Darwin's explanation is too puerile for any one professing to be a learned scientist to give. He says that the females preferred males with the least hair (?) until the hairy men gradually became extinct, because, naturally, under such a regime, the hairy men would die off, and, finally only hairless men to beget progeny would survive. What do sensible, serious students think of this "scientific" explanation? If we try to take this explanation seriously, we find that the science of phrenology teaches that females, as a rule, inherit the traits of their fathers, and males the traits of their mothers. Hence, not the males but the females would become hairless by this ridiculous process. How do evolutionists account for the hair left on the head and other parts of the body? Why do men have beard, while women and children do not? If the hair left on the body is vestigial, why is there no hair on the back, where it was most abundant on our brute ancestors? Even Wallace, an evolutionist of Darwin's day, who did not believe in the evolution of man, calls attention to the fact that even the so-called vestigial hair on the human form is entirely absent from the back, while it is very abundant and useful on the backs of the monkey family. If there was any good reason why the human brute should lose his hair, why for the same reason, did not other species of the monkey family lose their hair? Can it be explained by natural selection? Was the naked brute better fitted to survive than the hairy animal? Did man survive because he was naked, and the hairy brute perish? Evidently not, for the hairy brute still exists in great abundance.

The best way to get rid of the hair of the brute is for some reconstructing artist, like Prof. J. H. McGregor, to take it off. In a picture widely copied by books in favor of evolution, photographed from his "restorations," the pithecanthropus, the Neanderthal man, and the Cro-Magnon man are represented almost without hair on the body or even without beard. Only the Neanderthal man has a tiny Charlie Chaplin mustache. Their hair had not been combed for 1,000,000 years; yet we could not detect it. A sympathetic artist can make a "restoration" suit his fancy and support any theory.

If we are descended from simian stock, how did we come to lose our tails? Would not the same causes, if any, cause all the species to lose their tails? According to the laws of biometry, ought we not to find a retrogression of sections of the human race, who would sport simian tails and be clothed with simian hair? Or, could natural selection explain the loss of the tail on the ground that all the monkeys with tails died off, while the tailless ones survived, and developed into human beings? In that case, a tail must have been a fatal imperfection.

22
HYBRIDS

"Hybrids would seem to be nature's most available means of producing new species." Yet the sterility of hybrids defeats that possibility, and rebukes the untruthful claim of the formation of new species. Nature, with sword in hand, decrees the death of hybrids, lest they might produce a new species. Moses wrote the rigid unchanging law of nature, when he said that every living creature would bring forth "after its kind."

Species are immutable. One does not become another, or unite with another to produce a third. Dogs do not become cats, nor interbreed to produce another species. A few species, so nearly related that we can scarcely tell whether they are species or varieties, as the jackass and the mare, may have offspring, but the offspring are sterile. The zebra and the mare may produce a zebulon, which is likewise sterile. And so with the offspring of other groups intermediate between species and varieties. A human being and ape can not beget an ape-human, showing that they are not even nearly related species.

If evolution be true, we would expect a frequent interbreeding and interchanging of species. Even Darwin admitted that species are immutable. God declared it in his word, and stamps it indelibly on every species. "And God said, 'Let the earth bring forth the living creature after its kind, cattle, and creeping thing, and beast of the earth, after its kind'."-Gen. 1:24. How did Moses know this great truth, unless he was told by inspiration of God?

Even plant-hybrids are not permanent. Darwin himself says: "But plants not propagated by seed, are of little importance to us, for their endurance is only temporary."

Even if it could be proven that species, like varieties, are formed by development, it does not follow that genera and families and classes are so developed. But it has not been proved that a single species has been added by development, much less orders, families and genera. Evolution must account for every division and sub-division to plant and animal life. Darwin answers the objection to the sterility of hybrids by saying, "We do not know." "But why," he says, "in the case of distinct species, the sexual elements should so generally have become more or less modified, leading to their mutual infertility, we do not know." But God knows.

23
THE INSTINCT OF ANIMALS

The instinct of animals is not due to their own intelligence. It is unerring, unchangeable, without improvement or deterioration. It implies knowledge and wisdom of the highest order. It is beyond the wisdom of man. It comes direct from God. It is not learned nor gained by experience. It is found in many species of animals, and even in a child, until knowledge and reason make it unnecessary.

One of the most familiar illustrations is the instinct of the honey bee. It builds its cells in exact geometric form and we compute, by Calculus, that the form it uses produces the greatest capacity in proportion to the amount of material used. Who taught the bee to build its cell, displaying greater knowledge than that of many a college graduate? Darwin says (Origin of Species), "It can be clearly shown that the most wonderful instincts with which we are acquainted, namely those of the honey bee, could not possibly have been acquired by habit." We quote from Granville's Calculus, p. 119: "We know that the shape of a bee cell is hexagonal, giving a certain capacity for honey with the greatest possible economy of wax." This is demonstrated by the solution of a problem in this same Calculus. Darwin again says (Origin of Species, vol. I, p. 342), "We hear from mathematicians, that bees have practically solved a recondite problem, and have made their cells of the proper shape to hold the greatest possible amount of honey, with the least possible consumption of precious wax in their construction. It has been remarked that a skilful workman, with fitting tools and measures, would find it very difficult to make cells of wax of the true form, though this is effected by a crowd of bees, working in a dark room. Each cell, as is well known, is a hexagonal prism, with the basal edges of its six sides, beveled so as to join an inverted pyramid of three rhombs. These rhombs have certain angles, and the three which form the pyramidal base of a single cell on one side of the comb, enter into the composition of the bases of the three adjoining cells on the opposite side."

Can any one suggest an improvement or show an imperfection? If this intelligence is the bee's own, which is far superior to that of the ape, why did not the bee develop a human brain?

Yet in spite of Darwin's admission, he labors hard to show that "There is no real difficulty under changing conditions of life, in natural selection accumulating to any extent slight modifications of instinct which are in any way useful"! How could the working bee conserve the gains accumulated by experience or habit? The drone is the father and the queen is the mother of the sterile female working bee. Neither parent knows how to build a cell. How could they transmit their knowledge or their habits to the working bee? Every new swarm of bees would not know how to build their cells. There is no improvement from generation to generation. Even if instinct in other animals could be accounted for, evolution can not account for the instinct of the working bees, since they are not descendants of other working bees, from which they might inherit habits or instinct.

Is not the instinct of the bee the intelligence of God, disproving the heresy of an absentee God? Here again we get a glimpse of the unerring wisdom of God.

The immoveable oyster, the bee alive with divine intelligence, and the sterile progeny of the jackass, are enough to upset the whole theory of evolution.

24
SPECIAL CREATION: GEN. I

Evolution can not be true, because it contradicts the inspired word of God. We do not speak arbitrarily and say, without proof, that whatever contradicts the revealed word of God can not be true, although such an attitude could be easily defended. Disregarding all the many other cogent and legitimate arguments in support of a divine revelation, we will appeal to the remarkable harmony between the story of Creation in Genesis and the modern sciences. This could not be, if God had not revealed to Moses the story of creation. Moses personally knew nothing revealed by the sciences of today. And the man of that day who would invent the story of creation, would be sure to conflict with one or more of the following modern sciences: geology, astronomy, zoology, biology, geography, chemistry, physics, anatomy, philology, archaeology, history, ethics, religion, etc. There is not one chance in a million that a writer of a fictitious account would not have run amuck among many of these sciences, if, like Moses, he had no personal knowledge of them.

Although the Babylonian account may have had some foundation in fact, from a tradition of a prior revelation, it plainly bears the marks of error. "The Babylonian stories of creation are full of grotesque and polytheistic ideas, while those of the Bible speak only of the one living and true God." "All things," the Babylonian legend says, "were produced at the first from Tiamat." "The gods came into being in long succession, but, at length, enmity arose between them and Tiamat, who created monsters to oppose them. Merodach, a solar deity, vanquished Tiamat, cut her body in two, and with one-half of it made a firmament supporting the upper waters in the sky, etc., etc." The Babylonian gods, like even those of the classics, were criminals fit only for prison or death.

Alfred Russell Wallace, who, with Darwin, devised the evolution theory, says: "There must have been three interpositions of a Divine and supernatural power to account for things as they are: *the agreement of science with Genesis is very striking*: There is a gulf between matter and nothing; one between life and the non-living; and a third between man and the lower creation; and science can not bridge them!"

This "striking agreement" between science and Genesis I, is shown by the fact that at least 11 great events are enumerated in the same order as claimed by modern science: 1. The earth was "waste and void"; 2. "Darkness was upon the face of the deep"; 3. Light appears; 4. A clearing expanse, or firmament; 5. The elevation of the land and the formation of the seas; 6. Grass, herbs and fruit trees appear; 7. The sun, moon and stars *appear*; 8. Marine animals were created; 9. "Winged fowls" were created; 10. Land animals were created; 11. Man was created.

The chance of guessing the exact order of these 11 great events is ascertained by the law of permutations-the product of the numbers from 1 to 11, which is 39,916,800. Therefore, Moses had one chance out of 39,916,800 to guess the correct order of these 11 great events, as revealed both by science and revelation. If, for example, the first 11 letters of the alphabet were arranged in some unknown miscellaneous order, any one would have but one chance out of 39,916,800 to guess the order. If Moses did not have the order revealed to him, he never could have guessed it. Therefore, he was inspired and was told the order.

This mathematical demonstration annihilates the contradicting theory of evolution. At once it proves that the account was divinely inspired, and man came by special creation and not by evolution. The fact that the language of Genesis is in remarkable harmony with all proven modern scientific theories, and manifestly confirmed by them, is a proof in favor of the creation story, decisive and final.

This harmony is manifest whether the Heb. *yom*, day, be taken to mean a long period, as advocated by many biblical scholars, or a literal day of 24 hours, followed, it may be, by years or ages of continuance of the work, before the next day's work of 24 hours began.

Believing that this interpretation does no violence to the text, and that it is especially in harmony with the statements in the fourth commandment and elsewhere in the Bible, it is here briefly presented as one interpretation, showing the marvelous harmony between revelation and the proven, and even the generally accepted, scientific theories. The stately procession of events is the same, no matter which interpretation is accepted, and doubtless will remain, even if both must yield to another and better interpretation. This majestic divine order, in harmony with both science and revelation, removes all doubt of special creation.

Another interpretation, advocated by many scholars, is that all geologic ages may have intervened during the time indicated between the 1st and 2nd verses of Gen. I.

The following is a possible, and, it would seem, a probable interpretation of the inspired creation story. The words of Scripture, whether from the American Revision, or marginal rendering of the original Hebrew, or other translation, are put in quotation marks:—

THE CREATION—GENERAL STATEMENT

"In the beginning God created the heavens and the earth," including the sun, moon and stars, and all other matter in any form.

DETAILED STATEMENT OF THE ORDER OF CREATION

"And the earth was waste and void," literally "desolation and emptiness." And, on account of the thick vapors in the hot atmosphere, "darkness was upon the face of the deep," and doubtless had been for ages.

"And the Spirit of God was brooding upon the face of the waters," and *perhaps* was calling into being the lowest forms of marine life.

The First Day's Work. Light Appears.

"And God said, 'Let the light appear'," through the thick vapors. And the light appeared, so that the day could now be distinguished from the night. "And there was evening, and there was morning, one day." This day did not need to be an age or even 24 hours for God's work. How long did it take light to appear? Many years, and even ages, may have followed between each day's work as the "days" were not necessarily consecutive, and it is not so stated.

Second Day's Work. A Clearing Expanse.

"And God said, 'Let there be a clearing expanse (called heaven) dividing the waters which were on the earth from the waters in the thick clouds above, firmly suspended in the air'." This may have continued a long time, though begun in 24 hours.

Third Day's Work. Land, sea and vegetation appear.

"And God said, 'Let the waters under the expanse be gathered together into one place (seas and oceans), and let the dry land appear'." The contraction of the cooling earth caused the elevation of the land, and the draining of the waters into the seas. The geologist Lyell says, "All land has been under water." Hitchcock says, "The surface of the globe has been a shoreless ocean." "And the earth brought forth grass, herb yielding seed after its kind, and tree bearing fruit, wherein is the seed thereof, after its kind." Though the sun was not yet visible on account of dense clouds and vapors, the warm, humid atmosphere was suitable for the grass, herbs, and fruit trees,—three great classes which represented the vegetable kingdom. Ages may have again intervened.

The Fourth Day's Work. Sun, moon
and stars made visible.

"And God said, 'Let lights be seen in the open expanse of heaven, to divide the day from the night; and let them be for signs and for seasons, and for days and years'." "And God made the two great lights to *appear*," since neither had been seen through the thick clouds, "the greater light to rule the day, and the lesser light to rule the night. He made the stars also to *appear*." Though created first, the stars would appear last. Ages more may have intervened.

The Fifth Day's Work. Animal life in sea and air.

"And God said, 'Let the waters swarm with swarms of living creatures, and let birds fly above the earth upon the face of the expanse of the heaven'." "And God created great sea monsters, and every living creature that moveth which the waters brought forth abundantly, after their kinds, and every winged fowl after its kind." Geology and Moses alike testify that swarms of animals filled the seas. The ages rolled on while they "filled the waters of the seas and fowl multiplied on the earth."

The Sixth Day's Work. The creation
of land-animals and man.

"And God said, 'Let the earth bring forth the living creature after its kind, cattle and creeping things, and beast of the earth after its kind'." The fifth day animals began to *swarm* the seas; the sixth day, to cover the land. "And God said, 'Let us make man in our image, after our likeness'," in "knowledge after the image of him that created him," (Col. 3:10) and "in righteousness and true holiness," (Eph. 4:24). Yet a professor in a great university was so dense as to insist that the Scriptures taught that the likeness was not in "knowledge, righteousness and true holiness," but in the bodily form. "So God created man in his own image, in the image of God created he him." The last of all creation as both revelation and science testify. The image is mental and moral and spiritual. No such image in any other species.

The body chosen was higher and better than the form of any animal. It resembles the bodies of mammals of the highest type. Why should it not? The vast number of animal species, of almost every conceivable size and shape, could not furnish a form so well adapted to the use of man as that which the Creator gave him. Would it have been better if man had been created in the form of a fish, a lizard, a serpent, a dog, or a horse, or a bird? How could the body have been created without bearing resemblance to some form of the million species of animals? A resemblance can be traced through the whole creation, the material as well as the animal, but it does not follow that one species is descended from another, but that there was one general plan, and one God. The existence of man, who can not be otherwise accounted for, proves the existence of the Creator.

25
ANALOGY; MATHEMATICS, LAWS

Analogy raises a presumption against evolution. Analogy is not a demonstration. It is an illustration that strengthens and confirms other arguments. Both the science of mathematics and all physical laws must have come into being in an instant of time. Evolution is not God's usual method of creation.

1. **Mathematics**.—There is no evolution in the science of mathematics. There is no change or growth or development. God is the author of all mathematical principles. The square described on the hypotenuse of a right-angled triangle is equal to the sum of the squares described on the other two sides, because he made it so. The circumference of a circle is approximately 3.1416 times the diameter because he made it so. The wonderful calculations by logarithms, whether by the common system with a base of 10, or the Napierian system with a base of 2.718 a decimal that never terminates, are possible and reliable only because God made them so. Think what great intelligence is required by the Napierian system, to raise a decimal that never terminates, to a decimal power that never terminates, in order to produce an integral number. Yet God has computed instantaneously every table of logarithms, and every other mathematical table,—no matter how difficult. Thus we have positive proof of the presence everywhere of a great intelligent Being, and we catch a glimpse of that mind that must be infinite. He created the whole system of mathematics, vast beyond our comprehension, at once. A part could not exist without the whole. No growth; no change; no evolution; no improvement, because the whole system was perfect from the first. Reasoning from analogy, is it not reasonable to say that the God who flashed upon the whole universe, the limitless system of mathematics in an instant, also created man as Moses said? Analogy supports the doctrine of the special creation of man in a day.

The great system of mathematics which could not exist without a creator, is so extensive that 40 units are taught in a single university. New subjects are added, new text books written, new formulas devised, new principles demonstrated,—and the subject is by no means exhausted. He, by whose will this fathomless science came into existence, knows more than all the mathematicians of the past, present and future, and possibly all the evolutionists of the world.

2. **Physical Laws.**—All physical laws, prevailing throughout the universe, came into being by the will of God, in an instant of time. No growth, no change, no development, no evolution. The presumption is that God created all things in a similar way. If it was wisest and best to bring into being the great science of mathematics and fix all physical laws,—all in a moment of time, why should he consume 60,000,000 or 500,000,000 years in bringing man into existence? Evolution is all out of harmony with God's other methods of work.

Gravitation was complete from the first. No growth; no evolution. The laws of light, heat, electricity, etc., remain unchanged. Light travels with the same unvarying velocity, as when, 60,000 years ago, it started from the distant star-cloud. Some estimate our universe to be 1,000,000 light years across. Yet in all these limitless reaches, the same perfect and complete laws prevail, touching light, heat, electricity, gravitation, etc. God makes no mistakes and no evolution is needed. Does not this furnish a presumption that God could and did create man complete and full grown with a wonderful body, and a soul in his own image?

In this discussion, we have spoken of the "laws" of nature, after common usage. But laws are only a record of God's acts. An unchangeable God makes unchangeable laws. There is a rigid fixity written over the face of nature. Every law and principle is complete and perfect and finished, and there is no room for evolution.

Matter did not create itself, nor evolute nor grow. It must have been created instantaneously by the power of God, whether in a nebulous condition or not. So enchanting is their theory, that many profess to believe that not only were all species of animals and plants evolved from a single germ, but that even matter itself was evolved out of nothing. This theory of evolution as wide as the universe, as ponderous as the stars, is supported only by the weak stork legs of wistful possibility.

26
DESPERATE ARGUMENTS

Many arguments gravely given in support of evolution, reveal a great poverty of facts and logic. An instantaneous photograph of an "infant, three weeks old, supporting its own weight for over two minutes," is given by Romanes as a proof that man is descended from a simian (ape-like) ancestor. As this same picture is widely copied in evolution text books, they must have failed to get the picture of any other infant performing a like feat. Just how this affords any convincing proof that man is a monkey, we leave the reader to figure out. Our attention is called to the way this child and another child, whose picture is likewise generally copied, hold their feet (like monkeys climbing trees) showing they are little monkeys. Though we fail to see the force of this argument, it must be among their best from the emphasis they give it. Prof. H. H. Newman, of Chicago University, a leading evolutionist actually writes as follows, (Readings): "The common cotton-tail rabbit raises its white tail when it runs. This is interpreted [by whom, evolutionists or rabbits?] as a signal of danger to other rabbits."

The following absurd speculation, by a lecturer in the "University Extension Course," was printed in the Philadelphia Bulletin: "Evidence that early man climbed trees with their feet lies in the way we wear the heels of our shoes,—more at the outside. A baby can wiggle its big toe without wiggling its other toes,—an indication that it once used its big toe in climbing trees. We often dream of falling. Those who fell out of the trees some 50,000 years ago and were killed, of course, had no descendants (?) So those who fell and were not hurt, of course, lived, and so we are never hurt in our dreams of falling"! While we read these feeble arguments, which the newspapers would call piffle, how can we escape the conviction that evolution is in desperate need of argument? Imagine the Copernican theory relying on such piffle for support. Is there a freak idea without a freak professor to support it?

27
TWENTY OBJECTIONS ADMITTED

Evolutionists themselves, even including Darwin, admit as many as 20 objections to his theory. Darwin states the first four and Prof. V. L. Kellogg sums up the remaining 16 on pp. 247-52 of "Readings in Evolution." Among them are:—

1. There must have been innumerable transitional forms in the formation of new species. No convincing evidence of these missing links exists.

2. Natural selection can not account for the instinct of animals such as that of the honey bee, "which has practically anticipated the discoveries of profound mathematicians."

4. The offspring of such nearly related species as can be crossed are sterile, showing that nature discourages and in no wise encourages the formation of new species.

5. The changes resulting from the use and disuse of organs are not inherited.

6. Since Darwinism eliminates design, it is only the exploded ancient heathen doctrine of chance.

7. Variation is so slight as to be imperceptible, and, therefore, cannot account for the "survival of the fittest." If the same progressive changes do not occur generally, if not universally, in the numbers of the same species in the same period, no new species can arise. Such general changes do not occur.

8. Natural selection could not make use of initial slight changes. "What would be the advantage of the first few hairs of a mammal, or the first steps toward feathers in a bird, when these creatures were beginning to diverge from their reptilian ancestors?"

9. Even if Darwinism should explain the *survival* of the fittest, it does not explain the *arrival* of the fittest, which is far more important.

10. Darwin says, "I am convinced that natural selection has been the most important but not the exclusive means of modification." Many scientists think it of very little importance, and that it is not true.

11. "The fluctuating variations of Darwinism are *quantitative*, or plus and minus variations; whereas, the differences between species are *qualitative*." Growth and development in one species does not produce a new species, which must be of a different kind. Miles Darden, of Tenn., was 90 inches tall, and weighed 1000 pounds, but remained a member of the human species, though he was as high and heavy as a horse. So did the giant Posius, over 10 feet tall, who lived in the days of Augustus.

12. "There is a growing skepticism on the part of biologists as to the extreme fierceness of the struggle for existence and of the consequent rigor of selection." Overproduction and shortage of space and food might sometime be a factor of importance, but has it been so in the past? Has it affected the human race?

13. Darwin proposed the theory of gemmules. Prof. H. H. Newman says, "This theory was not satisfactory even to Darwin and is now only of historical interest."

14. Darwin's subsidiary theory of sexual selection has also been rejected by scientists as worthless.

In view of these and other objections, is it any wonder that Darwin's theory has been so largely rejected by the scientific world?

And is it not amazing that self-styled "scientists" hold on to their precious theory of evolution, as if these objections had no weight? They can not save evolution even by rejecting Darwinism.

28
SCIENTISTS CONDEMN EVOLUTION

Dr. Etheridge, famous fossilologist of the British Museum, one of the highest authorities in the world, said:—"Nine-tenths of the talk of evolutionists is sheer nonsense, not founded on observation and wholly unsupported by facts. This museum is full of proofs of the utter falsity of their views. In all this great museum, there is not a particle of evidence of the transmutation of species." Is a man in that position not a credible witness?

Prof. Beale, of King's College, London, a distinguished physiologist, said: "There is no evidence that man has descended from, or is, or was, in any way specially related to, any other organism in nature, through evolution, or by any other process. In support of all naturalistic conjectures concerning man's origin, there is not, at this time, a *shadow of scientific evidence*."

Prof. Virchow, of Berlin, a naturalist of world wide fame, said: "The attempt to find the transition from the animal to man has ended in total failure. The middle link has not been found and never will be. Evolution is all nonsense. It can not be proved by science that man descended from the ape or from any other animal."

Prof. Fleishman, of Erlangen, who once accepted Darwinism, but after further investigation repudiated it, said: "The Darwinian theory of descent has not a single fact to confirm it, in the realm of nature. It is not the result of scientific research, but is purely the product of the imagination."

Prof. Agassiz, one of the greatest scientists of any age, said: "The theory [of the transmutation of species] is a scientific mistake, untrue in its facts, unscientific in its method, and mischievous in its tendency.... There is not a fact known to science, tending to show that a single kind has ever been transmuted into any other."

Dr. W. H. Thompson, former president of N. Y. Academy of Medicine, said: "The Darwinian theory is now rejected by the majority of biologists, as absurdly inadequate. It is absurd to rank man among the animals. His so called fellow animals, the primates—gorilla, orang and chimpanzee—can do nothing truly human."

Sir William Dawson, an eminent geologist, of Canada, said: "The record of the rocks is decidedly against evolutionists, especially in the abrupt appearance of new forms under specific types, and without apparent predecessors.... Paleontology furnishes no evidence as to the actual transformation of one species into another. No such case is certainly known. Nothing is known about the origin of man except what is told in Scripture."

The foremost evolutionists, Spencer, Huxley and Romanes, before their death, repudiated Darwinism. Haeckel alone supported the theory and that by forged evidence.

Dr. St. George Mivert, late professor of biology in the University College of Kensington, calls Darwinism a "puerile hypothesis."

Dr. James Orr, of Edinburg University, says: "The greatest scientists and theologians of Europe are now pronouncing Darwinism to be absolutely dead."

Dr. Traas, a famous palaeontologist, concludes: "The idea that mankind is descended from any simian species whatever, is certainly the most foolish ever put forth by a man writing on the history of man." Does this apply to H. G. Wells?

Dr. N. S. Shaler, professor of Geol., in Harvard University, said: "It is not yet proved that a single species of the two or three millions, now inhabiting the earth had been established solely or mainly, by the operation of natural selection."

Prof. Haeckel, a most extreme evolutionist, confesses: "Most modern investigators of science have come to the conclusion that the doctrine of evolution, and particularly Darwinism, is an error, and can not be maintained."

Prof. Huxley, said that evolution is "not proved and not provable."

Sir Charles Bell, Prof, of the University College of London, says: "Everything declares the species to have their origin in a distinct creation, not in a gradual variation from some original type."

These testimonies of scientists of the first rank are a part of a large number. Many of them and many more, are given in Prof. Townsend's "Collapse of Evolution," McCann's "God or Gorilla," Philip Mauro's "Evolution At the Bar," and other anti-evolution books. Alfred McCann, in his great work, "God or Gorilla," mentions 20 of the most prominent scholars, who do not accept Darwinism. Yet they say, "All scholars accept evolution"!!

UNSOLICITED TESTIMONIALS

Agents for this 20,000 edition may show these selections, culled from a mass of warm world-wide testimonials, by able critics, authors, professors, editors, magazines, reviews, governors of states, and rulers of nations. "Unanswerable;" "an absolute demonstration;" "masterful;" "true to title;" "clear and convincing;" "scholarly and logical;" "timely;" "terse;" "interesting;" "best I ever read;" "costs $1, worth $5;" "fully disproves evolution;" also:—

"I finished your book today at two sittings. It is the most effective polemic on the subject, I have yet seen. You have marshalled the evidence of mathematics against the delusion of man's descent from brute ancestry, with telling effect."—PHILIP MAURO, Noted Attorney and Author.

"Evolution Disproved is not only a strong book from the scientific and argumentative viewpoint, but is also unique in many ways. We wish everybody would and could read it, especially those who are enamored with Evolution."—PROF. L. S. KEYSER, D.D., in the Bible Champion.

"Evolution Disproved is a sober, fully sustained and very remarkable book vindicating its title. It surely is one of the most conclusive of books, tearing to shreds Evolution pretensions. Absolutely unanswerable; in the very front rank of masterly books."—THE METHODIST.

"I have, for a third of a century, made Evolution a study, but Evolution Disproved really refutes the fallacy more completely than any other that I have seen. Some rich man should give it to 20,000,000 families."—REV. C. W. BIBB, N.Y.

"You certainly have given a masterful treatment of this subject."—C. L. HUSTON, Chairman Com. on Evangelism, Pres. Church, U.S.A.

"Interessante" (French).—President of the Swiss Confederation.

"Filled with valuable matter systematically arranged; cogent."—S.S. TIMES, Philadelphia.

"He shows the evolution of the soul to be impossible."—W. R. MOODY, in Record of Christian Work.

"Unexcelled for brevity, clarity and intensity. A compendium of facts."—W.C.F.A., which accordingly rewarded the author with honorary membership.

"The arguments amount to a demonstration."—LUTHERAN, Phila.

"The greatest book of its kind."—PROF. M. F. LARKIN, head of the International Textbook Co., Scranton, Pa.

"A very informing book."—Bp. NUELSEN'S, Sec., Zurich.

"A most remarkable book."—THE LUTHERANEREN (Danish)

"A vigorous book; a lively volume."—BELFAST (Ireland) NEWS.

"A strong argument."—GUERNSEY PRESS, Eng.

"A very remarkable and provocative book; shows patent evidence of large research and shrewd thinking."—COURIER, Dundee, Scotland.

"I congratulate you on this scientific work so full of thought."—H. SEIPEL, Chancellor of Austria,

"An excellent book."—Librarian of Ravenna University, Italy.

"An interesting attack on evolution."—Teachers World, London, Eng.

"A very excellent book."—REV. D. D. MARSH, Ont., Can.

"The best I ever saw."—R. A. McKINNEY, G. A. Com. of 100.

"Irrefutable; displays unusual information."—Dr. D.S. Clark. Phila.

"He writes from a new angle with great ability."—Luth. Church Her.

"Should do much good."—REV. F. HAMILTON, Pyongyang, Korea.

"I count your book a remarkably strong one. It clearly disproves every claim of Darwinism."—DR. H. B. RILEY, President W.C.F.A.

"Of all books against evolution, the most unique. Its arguments are effective and deadly, cumulative and convincing."—Bibliotheca Sacra.

"Our first order, 60 copies."—BIBLE UNION, Cape Town, S. Africa.

"Thanks" for EVOLUTION DISPROVED have been received from HUNDREDS of foreign librarians and national rulers. Write what YOU think!

PART TWO
EVIDENCE ANSWERED

29
PALEONTOLOGY

1. The Pithecanthropus, which is a high sounding name for an ape-man (from Grk. pithekos, ape, and anthropos, man) was found by Dr. Dubois, an ardent evolutionist, in 1892, in Trinil in the island of Java. It lived, it is said, 750,000 years ago. He found, buried in the Pleistocene beds, 40 feet below the surface in the sand, *the upper portion of a skull, a tooth and a thigh bone*. "It was fortunate," says Dr. Chapin, "that the most distinctive portions of the human (sic) frame should have been preserved, because from these specimens, we are able to reconstruct (?) the being, and to say with assurance (!) that his walk was erect in manlike posture, that he had mental power considerably above the ape, (it will not do to be too definite) and his powers of speech were somewhat limited. (A string of guesses wholly unwarranted.) This man stood half way between the anthropoid and the existing men." —Social Evolution, p. 61.

A high authority declares,— "Shortly after this discovery, 24 of the most eminent scientists of Europe met. Ten said that the bones belonged to an ape; 7, to a man; and 7 (less than one-third) said they were a missing link." Some of the most eminent scientists say that some of the bones belong to a man, and some to an ape, baboon, or monkey. The great Prof. Virchow says: "There is no evidence at all that these bones were parts of the same creature." But such adverse opinions do not weigh much with modern evolutionists determined to win at all hazards.

The small section of the brain pan, weighing but a few ounces, was found about 50 feet from the thigh bone. One tooth was found 3 feet from the fragment of skull, and one near the thigh bone, *50 feet away*. Since the small section of the brain pan belonged to a chimpanzee, and the thigh bone is that of a man, is it likely that these scattered bones belonged to the same

The Evolution Of Man Scientifically Disproved | *93*

creature? Even if they did, is it likely that these bones would be preserved in the sand 750,000 years, or even 375,000 years according to a later estimate? We know that petrified skeletons, encased in rock, may be millions of years old, but where are the unpetrified skeletons of men who lived even 5,000 years ago? If unpetrified skeletons could last 750,000 years, there would be millions of them. Without a doubt, this skull of a chimpanzee, and femur of a man, belong to a modern beast and a modern man, buried by floods or earthquakes, or some other convulsion of nature, or by slow accumulations. It is said that the Jerusalem of Christ's day is buried 20 feet under the surface, by the quiet accretions of the dust of 1900 years. Rome also has been covered up in recent centuries. It would be easy for 40 feet of sand to accumulate over the bones of a modern man or chimpanzee in a valley, in a few centuries, if 20 feet of dust accumulated on the mountain city of Jerusalem in 1900 years.

Elsewhere we have shown that an ape-man with a cranium of two-thirds normal capacity must have lived at least 20,000,000 years ago,—one third the period of animal existence; or even 166,666,666 years ago, if we accept a later claim that life has existed 500,000,000 years. It is absolutely impossible that a normal creature of the alleged mental capacity could have lived 750,000 years ago, much less 375,000, according to a later estimate cutting in two the first one. But the quickest way to disprove these wild guesses is to check them up by a mathematical test. If these bones are normal, such an ape-man could not have lived at the time assigned. If they are not normal, they prove nothing whatever for evolution. They can be duplicated now.

We are asked to believe that these scattered bones,—some the bones of a modern brute, some the bones of a modern man—were preserved in the sand 750,000 years and belonged to an ancestor of the human race, while of the millions of his generation and of the generations following for many thousands of years, we have not a trace! We are asked upon such a flimsy pretext to accept a theory, unsupported by a single compelling argument, and irreconcilable with numerous facts,—a theory which takes away man's hope of immortality, destroys faith in God and his inspired word, and in the Christian religion itself. There is a limit. How much more truthful and majestic is Gen. 1:27: "And God created man in his own image, in the image of God created he him."

One distinguished evolutionist has said, "We might as well be made out of monkey as out of mud. It is mud or monkey." Most of us would retort, "I would rather be created a human being out of the filthiest mud by Almighty God than owe my existence to the brainiest monkey that ever lived." Please note, "The Lord God formed man of the *dust* of the ground," not *mud*. The evolutionists are as wild in their exegesis as in their guesses.

2. The Heidelberg Jaw. The second relic, in the order of time, relied upon by the evolutionists to prove the brute origin of man, is a "human jaw of great antiquity, discovered in the *sands* of the Mauer River, near Heidelberg." Hence, it is called the Mauer jaw, or the Heidelberg Jaw, or Heidelberg man, or the high sounding Latin name of Homo Heidelbergensis. It needs all the names that can be given to it, to elevate it to the dignity of an ancestor. "This jaw was found in undisturbed stratified *sand*, (sand again) at the depth of about 69 feet from the summit of the deposit." Dr. Schoetensack, the discoverer, says, "Had the teeth been absent, it would have been impossible to diagnose it as human."

They say it is 700,000 years old, preserved in sand. A later estimate says 375,000 years. (Any wild guess will do.) It resembles the jaw of an ape, and the tooth of a man. Was it not likely the abnormal jaw of a modern man, in historic time swept into the sands by the freshets and floods of a few centuries? It is only fair to say that many scientists of the evolutionary school, do not believe the Heidelberg man an ancestor of our race. "These remains," says one, "show no trace of being intermediate between man and the anthropoid ape." Some claim it a connecting link. Others deny it. Some say the find is of the utmost value; others say it is worthless. All are guesses, wild guesses at that. They hopefully reach out their hands in the night, and gather nothing but handfuls of darkness.

Since a modern Eskimo skull has been shown by a distinguished scientist to have the same appearance and peculiarities as the Heidelberg jaw, it is easy to believe that this jaw can be duplicated in many graveyards. Greater abnormalities, in great numbers, can be found in the skeletons of modern man. Without doubt, this jaw belongs to modern man, and has no evidential value at all in favor of evolution.

We count these relics normal, in our arguments, because evolutionists do. If they are not normal, they are the remains of modern man and brutes and their whole argument falls to the ground.

3. **The Piltdown Man (or Fake).** The next fragments of bones, in chronological order, upon which evolutionists rely to prove their impossible theory, has been called the Piltdown man. It has been more truthfully called the Piltdown fake. Dr. Chapin gravely tells us (Social Evolution, p. 67): "During the years 1912, a series of fragments of a human skull and a jaw bone were found associated with eolithic implements and the bones of extinct mammals in Pleistocene deposits on a plateau, 80 feet above the river bed, at Piltdown, Fletching, Sussex, Eng.....The remains were of great importance. The discoverers regard this relic as a specimen of a distinct genus of the human species and it has been called Eoanthropus Dawsoni.

This extinct man lived in Europe hundreds of thousands of years ago." We have passed over 200,000 to 300,000 years since the Heidelberg man, that have not yielded a scrap of bone, though according to the theory, countless millions of ape-men must have lived in various stages of development, in that great stretch of time. Why were not some of them preserved? Simply because there were no ape-men. There are countless relics of apes, but none of ape-men. Even Wells says: "At a great open-air camp at Solutre, where they seem to have had annual gatherings for many centuries, it is estimated there are the bones of 100,000 horses." Would we not expect as many bones of ape-men? While Wells says the bones of 100,000 horses were found in a single locality, Dr. Ales Hrdlicka says that the bones of 200,000 prehistoric horses were found in another place. Why should we not find, for the same reason, the bones of millions of ape-men and ape-women in 750,000 years? Instead of millions we have the alleged fragments of 4, all of which are of a very doubtful character.

The bones of this precious Piltdown find consisted, at first, of a *piece of the jaw bone, another small piece of bone from the skull,* and a canine tooth, which the zealous evolutionists located in the lower right jaw, when it belonged in the upper left; later, two molar teeth and two nasal bones,—scarcely a double hand full in all. An ape-man was "reconstructed" made to look like an ape-man, according to the fancy of the artist. The artist can create an ape-man, even if God could not create a real man! But scientists said the teeth did not belong to the same skull, and the jaw could not be associated with the same skull. Ales Hrdlicka says, "The jaw and the tooth belong to a fossil chimpanzee." Conscientious scientists said that the pieces of the jaw and skull could not belong to the same individual. They constructed a scarecrow from the bones of an ape and of a man, and offer this, without the batting of an eye, as a scientific proof of the antiquity of man. The great anthropologist of world-wide reputation, Prof. Virchow, said: "In vain have Darwin's adherents sought for connecting links which should connect man with the monkey. *Not a single one has been found.* This so-called pro-anthropus, which is supposed to represent this connecting link, has not appeared. No true scientist claims to have seen him." Sir Ray Lancaster, writing to H. G. Wells, concerning the Piltdown find, says, "We are stumped and baffled." Yet in spite of all this, nearly 1,000,000 persons annually pass through the American Museum of Natural History in New York, and view the "reconstruction" according to the artist's fancy, of the pithecanthropus, the Heidelberg man, the Piltdown man, and the Neanderthal man, the "ancestors of the human race;" and the multitude of high school students and teachers, as well as the general public, are not told how dubious and unscientific the representation is.

The brain capacity of the Piltdown individual (man or ape) is set down by his discoverers at 1070 c.c., which is 28 2/3% short of the normal skull capacity, 1500 c.c. Therefore, he must have lived 17,200,000 years ago, if we accept the estimate of 60,000,000 years since life began; or 143,333,333 years ago, if we accept the later guess of 500,000,000 years. It could not have lived near the time assigned. In short, no guess of the origin of man that differs materially from the time assigned in the word of God, can be harmonized with the facts.

4. The Neanderthal Man. The next slender prop is the Neanderthal man, claimed to be 40,000 to 50,000 years old, although we are told that that is very uncertain.

Dr. Chapin says, "The first important discovery of the existence of an early example of mankind differing markedly from any living (?) and of a decidedly lower type, was made in 1857, when a part of a skull was found in a cave near Dusseldorf, Germany. The bones consisted of the upper portion of a cranium, remarkable for its flat retreating curve, the upper arm and thigh bones, a collar bone, and rib fragments." From these fragments, an ape-man has been created (by the artist), about 5 ft. 3 in. high, strong, fierce in look, and having other characteristics created by the artist.

Dr. Osborn assigns to the Neanderthal skull a capacity of 1408 c.c., which would indicate that he lived 3,680,000 years ago, if life began 60,000,000 years ago; or 30,666,666 years ago, if life began 500,000,000 years ago.

From the first, many naturalists claimed that these bones belonged to an abnormal specimen of humanity. They can be easily duplicated. Naturalists have maintained many divergent opinions: an idiot, an early German, a Cossack, a European of various other nationalities, a Mongolian, a primitive ape-man, an ancestor of modern man, and an impossible ancestor of man. Not very reliable evidence to support the stupendous scheme of evolution!

Now these four finds are the weak props supporting the desperate claim of the brute origin of man. Dr. Chapin says (Social Evolution, p. 68): "Other skulls and bone parts of prehistoric man have been found, and preserved in museums, but the specimens described (the four above mentioned) are sufficient to illustrate *the type of evidence* they constitute." The later finds measuring close to normal capacity, doubtless are the bones of the descendants of Adam. Even by the admission of this text-book author, the evidence from other remains is no more convincing than that from these four types. Some evolutionists say that the pithecanthropus, the Heidelberg man, the Piltdown man, and the Neanderthal man, form an unbroken line of descent from the ape, each in turn becoming less like the ape, and more like man. Others claim that the pithecanthropus was the end of a special branch of the apes; the Heidelberg man the last of another extinct branch; the Piltdown man and the Neanderthal man, likewise the last of other extinct species. In this case, all four finds have no evidential value whatever. All these confusing guesses from evidence so scant and uncertain, stamp evolution a "science falsely so called."

If these branches, species, or races of ape-like creatures ended, as claimed, in the age to which these alleged remains belonged, they could not have been the ancestors of the human race, and these alleged links were not links at all. Some evolutionists say that the Neanderthal race became extinct 25,000 years ago. If so, they were not our ancestors. We are curious to know what caused the extinction of all these races. Prof. R. S. Lull confesses, "However we account for it, the fact remains that ancient men are *rare*." Most unbiased students would say such men never existed. The entire absence of human remains during the 750,000 years and more is a demonstration against the brute origin of man, and a proof of special creation.

It will be remembered that there is no complete skeleton among all the remains, nor enough parts to make one altogether, nor to make any large part of a skeleton,—not even an entire skull. What bones are found are not joined together, and some of them scattered so widely apart, that no one can be certain they belong to the same individual. Some of the bones belong to an ape, and some to man,—doubtless modern man. Ardent evolutionists, with a zeal worthy of a better cause, have taken a fractional bone of a man, and a bone of an ape, and fashioned a composite being, and called it an ape-man, and their ancestor.

Every one of these finds is disputed by scientists, and even by evolutionists. And all these doubtful relics would not fill a small market basket. Yet some are ready to say that evolution is no longer a guess or a theory, but a proven fact. Text books like Chapin's Social Evolution are placed in the hands of pupils giving only the arguments in favor, and the student, even if disposed to question this flimsy and unsupported theory, is helpless in the hands of an adroit professor. Dr. Gruenberg's high school text book teaches that man is descended from the pithecanthropus, the Heidelberg, the Piltdown and the Neanderthal man, without the slightest intimation that such descent is at all disputed or questioned. What right has anyone to teach this false and unproved theory as the truth?

30
CONFESSED COLLAPSE OF "PROOF"

The claim that the pithecanthropus, the Heidelberg man, the Piltdown man, and the Neanderthal man, were the ancestors of man, collapses under the admissions of evolutionists themselves. The eminent Wassman says: "There are numerous fossils of apes, the remains of which are buried in the various strata from the lower Eocene to the close of the alluvial epoch, but *not one connecting link* has been found between their hypothetical ancestral forms and man at the present time. The whole hypothetical pedigree of man is *not supported by a single fossil genus or a single fossil species*" (all italics ours). Darwin says: "When we descend to details, *we can prove that not one species has changed.*" How, then, can man be descended from the brute?

Even H. G. Wells, who seems ready to endorse the most extravagant views, says (Outline of History, p. 69), "We can not say that it (the pithecanthropus) is a direct human ancestor." On p. 116, is a "Diagram of the Relationship of Human Races," showing that neither the pithecanthropus, the Heidelberg man, the Piltdown man, nor the Neanderthal man, could have been an ancestor of the human race, because each were the last of their species, and therefore had no descendants.

Dr. Keith, a London evolutionist, says that the Piltdown man is not an ancestor of man, much less an intermediate between the Heidelberg man and the Neanderthal man. Sir Ray Lancaster confesses he is "baffled and stumped" as to the Piltdown man. Dr. Keith says the "Neanderthal man was not quite of our species."

Dr. Osborn says that the Heidelberg man "shows no trace of being intermediate between man and the anthropoid ape." Again, speaking of the teeth of the St. Brelade man, Dr. Osborn says, "This special feature alone would exclude the Neanderthals from the ancestry of the higher races."

Prof. R. S. Lull says, "Certain authorities have tried to prove that the pithecanthropus is nothing but a large gibbon, but the weight of authority considers it prehuman, though not in the line of direct development in humanity."

Prof. Cope, a distinguished anatomist, says, "The femur [of the pithecanthropus] is that of a man, it is in no sense a connecting link."

In his "Men of the Old Stone Age," Dr. Osborn puts the pithecanthropus, the Heidelberg man, the Piltdown man, and the Neanderthal man, on limbs which *terminate abruptly as extinct races*. They can, in no sense, then, be the ancestors of man, or connecting links. Why, then, do they cling so desperately to these alleged proofs, when they admit they have no evidential value? Only sheer desperation, just as a drowning man will clutch a straw.

Dr. W. E. Orchard says: "The remains bearing on this issue, which have been found are very few, and their *significance is hotly disputed by scientists themselves,—both their age, and whether they are human or animal, or mere abnormalities.*"

Since these four creatures (of the evolutionists) can not be the ancestors of the human race, where are their descendants? Evolutionists are obliged to say they were the last of their kind. Strange! But there is no other way of escape.

Prof. Bronco, of the Geological and Palaeontological Institute of Berlin University, says, "*Man appeared suddenly in the Quaternary period. Palaeontology tells us nothing on the subject,—it knows nothing of the ancestors of man.*"

As fossils must be imbedded in rock, there is not a single fossil of an ape-man in the world.

31
PICTURES IN CAVERNS

To bolster up the hypothesis, that some of the scraps of bones belonged to ape-men; who lived about 50,000 years ago, we are told that, in many caverns there are paintings of animals, some of which are extinct, proving that the artists were ape-men of advancing intellect, living in that day. These drawings are rude, and inexact, and the resemblance to extinct animals rather fanciful. If the writer were to try to draw a picture of a horse on the stone walls of a dark cavern, with no light, it would be just as likely to resemble an extinct animal, or possibly an animal that never did live and never will. Many of the paintings are found in the depths of unlit caverns, often difficult of access. How could they paint any picture in the dark, when even fire was unknown, and the torch and lamp-wick had not yet been invented? And how could they make a ladder, or erect scaffolding of any sort in that rude age, before there were inventions of any kind? Yet they tell us that the frescoes on the ceiling of the dark cavern of Altamira, Spain, were made 25,000 to 50,000 years ago, when fire was unknown, and they ask us to believe that several colors are used, brown, red, black, yellow, and white; and that these drawings and colors have remained undisturbed and unchanged through these long ages. Is it easier to believe this, than to believe that these drawings were made by modern man, using modern inventions? A theory left to such support, must be poverty-stricken in argument indeed.

32
VESTIGIAL ORGANS

The claim is made that the so-called rudimentary organs in the human body such as the appendix, are the remnants of more complete organs inherited from our animal ancestors. It is a strange argument that a once complete and useful organ in our alleged animal ancestors, when it becomes atrophied in man, causes such an improvement and advance, as to cause man to survive, when his ancestors with more perfect organs became extinct. Man with less perfect organs became the dominant species. If the perfect organ were better than the rudimentary organ, how can man be the "survival of the fittest"? If rudimentary organs are a proof of descent from animals with more extensive, if not more perfect, organs, then both man and monkeys must be descended from the rat, which has the longest proportionate appendix of all. If unused muscles speak of our ancestry, the horse has the strongest claim to be our ancestor.

But many organs, such as "the thyroid gland, the thymus gland, and the pineal gland," formerly classified as rudimentary organs, are found to be very useful and necessary.

Physicians have found the appendix very useful in preventing constipation, which its removal usually increases. If we only knew enough, we would, no doubt, discover a beneficial use for all the so-called vestigial organs. Our ignorance is no argument against the wisdom of their creation. The claim that human hair is vestigial is spoiled by the fact that there is none on the back where most abundant on simians.

33
SEROLOGY, OR BLOOD TESTS

They tell us that the blood of a dog injected into the veins of a horse, will kill the horse, whereas the blood of a man injected into the veins of an ape results in very feeble reaction, which proves that the dog and the horse, they say, are not related by blood, while the man and the ape are so related. But a distinguished authority says, "The blood of the dog is poisonous to other animals, whilst, on the other hand, the blood and the blood serum of the *sheep, goat* and *horse*, have generally little effect on other animals *and on man*. It is for this reason that these animals and particularly the horse, are used in preparation of the serums employed in medicines."

It is also stated as a fact that mare's milk more nearly resembles human milk than that of any other animal save the ass, a nearly related species—to the mare, let us hope, not to us. Because of this resemblance, it is reported by Dr. Hutchinson that, "One of the large dairy companies in England now keeps a stock of milch asses for the purpose of supplying asses' milk for delicate human babes."

These well-known facts would prove the horse and the ass a nearer relative than the ape, since serums are not made from the blood of the ape. We prefer the innocent sheep to the ape as our near relative, and will allow the evolutionists to claim the goat.

Dr. W. W. Keen, Prof. Emeritus of Jefferson College, Phila., in his book, "I believe in God and in Evolution," on p. 48 says, "Here again you perceive such identity of function, that the thyroid gland of animals, when given as a remedy to man, performs precisely the same function as the human thyroid. Moreover, it is not the thyroid gland from the anthropoid apes that is used as a remedy but that from the more lowly sheep." Again the force of Dr. Keen's argument goes to prove, so far as it has any weight, that we have a nearer kinship to the sheep than the ape. Children are nourished by the milk of the cow, the ass and the goat, not of the ape. Vaccine matter is taken from the cow and serums from the horse, not from any species of monkey, to which we do not seem to be related at all.

The conclusions of the blood tests are unreliable and uncertain. W. B. Scott, an expert evolutionist, says, "It must not be supposed that there is any exact mathematical ratio between the degrees of relationship indicated by the blood tests, and those which are shown by anatomical and palaeontological evidence.... It could hardly be maintained that an ostrich and a parrot are more nearly allied than a wolf and a hyena, and yet that would be the inference from the blood tests."

Prof. Rossle, in 1905, according to McCann, presented evidence to show that the blood reaction does not in any manner indicate how closely any two animals are related; and that evidence based on resemblance of blood is not trustworthy in support of a common relationship. In many cases, transfusions of the human blood into apes have positive reactions. We do not make pets of the ape, baboon or chimpanzee, but of the dog whose traits are far more nearly human. If any brute ancestor is possible, have not the evolutionists guessed the wrong animal?

34
EMBRYOLOGY

Embryology, or the Recapitulation Theory, is the last, and perhaps the least important of the claims advanced in favor of evolution. It is claimed that the whole history of evolution is briefly repeated in the early stages of embryonic life. W. B. Scott, in the "Theory of Evolution," says, "Thirty years ago, the recapitulation theory was well nigh universally accepted. Nowadays it is very seriously questioned, and by some high authorities is altogether denied."

It is hard to see why the history of the species should be repeated by the embryo. It is difficult to crowd the history of ages into a few days or weeks. It must be enormously abbreviated. It is a physical impossibility. Changes caused by many environments must take place in the same environment, contradicting the theory of evolution. So many exceptions must be made that there can be no universal law. Such general similarity as we find in embryonic life, may be accounted for, on the ground that the Creator used one general plan with unlimited variation, never repeating himself so as to make two faces or two leaves or two grains of sand exactly alike.

"Embryology is an ancient manuscript with many of the sheets lost, others displaced, and with spurious passages interpolated by a later hand." It is hard to construct a syllogism, showing the force of the argument from Embryology. Try it.

Various other evolution arguments are answered in PART ONE, and completely refuted by UP-TO-DATE SCIENTIFIC FACTS. No one has yet noted an error, nor answered an argument. If all students, teachers, ministers, etc., had this book (pp. 116-7), evolutionists could no longer conceal the "unanswerable arguments," nor answer them by ridicule or abuse.

PART THREE
THE SOUL

35
THE ORIGIN OF THE SOUL

Evolution fails to account for the origin of the body of man. Still more emphatically, does it fail to account for the origin of the soul, or spiritual part of man. This is part of the stupendous task of evolution. Its advocates give it little or no attention. We are not surprised. If they *could* show the evolution of the human body *probable* or even *possible*, they can never account for the origin of the soul, save by creation of Almighty God. We can not release evolutionists upon the plea that they cannot account for the faculties and spiritual endowments of man. This is a confession of complete failure. Though invisible to the eye or the microscope, they are positive realities. They can not be dismissed with a wave of the hand or a gesture of contempt. We have a right to demand an explanation for every phenomenon connected with the body or soul of man. The task may be heavy, and even impossible, yet every hypothesis must bear every test or confess failure. They have undertaken to propose a scheme that will account for the origin of man, as he is, soul and body, and if they fail, the hypothesis fails.

How do we account for the existence of each individual soul? It can not be the product of the arrangement of the material of the brain, as the materialists do vainly teach. It can not be the product of evolution, nor a growth from the father or mother. The soul is not transmitted to be modified or changed. It is indivisible. The soul of the child is not a part of the soul of either parent. The parents suffer no mental loss from the new soul. It must be created before it can grow. God creates each soul without doubt, and so God created the souls of Adam and Eve. If creation is possible now, it was possible at the beginning of the race. If God creates the soul now, analogy teaches strongly the creation of the souls of Adam and Eve. If evolution be true, there was no creation in the past, and is none now. This is contradicted by the facts every day and every hour.

36
PERSONALITY

An evolutionist writes: "We do not undertake to account for personality." We reply, "That is a part of your problem. You have undertaken to solve the riddle of the universe by excluding all evidence of an existing and active God, and we can not release you because a feature of the problem may be unusually difficult or embarrassing, or even fatal to your theory. It is a fight to the death in the interest of truth; and we purpose to use every weapon of science against a theory so unscientific, so improbable, so far reaching, and so baneful in its effects. It takes faith, hope and comfort from the heart of the Christian, destroys belief in God, and sends multitudes to the lost world."

Personality is consciousness of individuality. When did personality begin? When did any members of the species become conscious of personality? When did they begin to realize and to say in thought, "I am a living being." What animals are conscious of personality? Any of our cousins of the monkey tribe? Is the horse conscious of personality, or the ox, the cat or the dog? If so, does the skunk have personality, the mouse, the flea, the worm, the tadpole, the microscopic animal? If so, do our other cousins have personality,—the trees, the vines, the flowers, the thorn and the brier, the cactus and the thistle, and the microscopic disease germs? If so, when did personality begin? With the first primordial germ? If so, were there two personalities when the germ split in two, and became two, animal and plant? You can not split a man up into two parts with a personality to each part. Personality is indivisible. It is a consciousness of that indivisibility. If personality began anywhere along the line, where, when, and how did it originate? Was it spontaneous, or by chance, or was it God-given? Beyond all question, it was the gift of an all-wise and all-powerful Creator, and in no sense the product of evolution. God made man a living soul.

But if no plant or animal ever had personality, when did man first become conscious of his individuality? There is no evidence, of course, but the evolutionist must produce it, or admit failure. The evolutionist is short on evidence but long on guesses that miss the mark.

If all animals and plants came from one germ, why do animals have the senses, sight, taste, touch, smell and hearing, while plants are utterly devoid of them? They had a nearly equal chance in the race. Why the great difference?

37
INTELLECT, EMOTIONS AND WILL

The activity and energy of the soul are shown in the intellect, the emotions and the will. What evidence of these do we find in the animal world? Do we find intellect in the lobster, emotions in a worm, or will in an oyster? Whence came these elements of spiritual strength? If developed by evolution, where, when, and how?

Have the most advanced species of animals an intellect? Do they have the emotions of love, hate, envy, pity, remorse or sympathy? Has a worm envy, a flea hate, a cat pity a hog remorse, or a horse sympathy? If these existed in so-called pre-historic man, when, where, and how did they begin? No one can answer, because there is not a trace of proof that they ever existed.

Will natural selection explain the development of the mental faculties? Was art developed because those who lacked it perished? Do we account for the musical faculty, because those who could not sing perished? Some still live who ought to be dead! Do we account for humor because they perished who could not crack a joke? Will all eventually perish but the Irish, who will survive by their wit? Is anything mentioned in science quite so ridiculous as natural selection?

Not an animal has a trace of wit, or humor, or pathos. Not an animal has ever laughed, or spoken, or sung. The silence of the ages disproves evolution.

38
ABSTRACT REASON

When did reason begin? Do we find it in any species of plant or animal life, save man? The highest order of animals can not reason enough to start a fire or replenish one. A dog, or a cat, or even a monkey, will enjoy the warmth from a fire but will not replenish it, although they may have seen it done many times. Animals may be taught many interesting tricks; many can imitate well. But they do not have the power of reflection or abstract reason. They live for the present. They have no plans for tomorrow,—-no purpose in life. They can not come to new conclusions. They can not add or subtract, multiply or divide. They can not even count. Some animals can solve very intricate problems by instinct, but instinct is the intelligence of God, and never could have come by evolution.

If reason came not from God, but from evolution, should we not expect it well developed in evolutionary man, since for the last 3,000,000 years he must have been 95 to 100 per cent, normal. If we grant the estimate of 500,000,000 years, he would have been 99.4% normal for the last 3,000,000 years. Would we not expect in that time a world of inventions and discoveries, even surpassing those of the last 100 years? The Chinese claim a multitude of inventions and a race so nearly normal as ape-men, ought to have invented language, writing, printing, the telegraph, phonograph, the wireless, the radio, television, and even greater wonders than in our age.

There is no trace of intelligence in man in all the 3,000,000 years, prior to Adam.

We should have many works excelling Homer's Iliad, Vergil's Aeneid, and Milton's Paradise Lost. We have no trace of a road, or a bridge, or a monument, like the pyramids. That no race of intelligent creatures ever lived prior to Adam is proven by lack of affirmative evidence. If it be true, as Romanes declared, that the power of abstract reason in all the species was only equal to that of a child 15 months old, then each species would possess less than one millionth of that.

39
CONSCIENCE

If the origin of the mental faculties can not be accounted for by evolution, much less can the moral faculty, the religious nature and spirituality be accounted for.

The most confirmed evolutionist will not claim that the tree or the vine or the rose, or perhaps any animal, has a conscience. If, however, conscience is a growth or development, why should it not exist in some measure in both the animal and the vegetable kingdoms? Has any brute any idea of right or wrong? Has a hog any idea of right or wrong, of justice or injustice? What animal has ever shown regret for a wrong, or approval of right in others? If conscience is a development within the reach of every species, many of the million or more, no doubt, would have shown some conscience long ago.

But if man developed conscience, why have not our near relatives of the monkey family developed a conscience? They had the same chance as man. Why should man have a conscience, and monkeys none?

Why is there no trace of conscience in the animal or vegetable kingdom? Because it is the gift of God.

What sign of regret, repentance, or remorse, do we find in the cat or the dog, the rat or the hog? If a bull gores a sheep to death, does he express regret? Is a horse sorry if he crushes to death a child or a chicken under his hoof? Can any animal be sorry for stealing food from another? Will it take any steps to undo the wrong?

Man, according to evolution, is a creature of environment. He is a victim of brute impulse. He has no conscience, no free will, he can commit no crime. Killing is not murder. It is not sin. Man can not be responsible. Without conscience, a victim of circumstances, rushed on into crime, sin, and injustice, responsible to no God!

The heart sickens at the brightest picture evolution can paint. The difficulty of showing the evolution of the body is insuperable, but the evolution of the soul, with all its mental, moral and spiritual equipment, is an absolute impossibility. Small wonder that evolutionists are unwilling to discuss the origin of the soul.

40
SPIRITUALITY

Does any plant or animal worship God? How much theology does a cow know? What does the horse think about God? What animal lives with an anxious desire to please God? How many are desirous of obeying God? How many species trust Him? How many love Him? How many pray to Him? How many praise Him for his goodness? Evidently no animal knows anything about God, or ever thinks of worshiping Him.

Man alone worships God. When did he begin? The idea of God seems to be in the hearts of all except the dupes of evolution, and the Bolshevists of Russia. The great problem to explain is how the worship of God began, and why man alone now worships Him.

Personality, reason, intellect, emotions, will, conscience, spirituality, and all the faculties and equipment of the soul, are naturally and easily explained upon the basis of creation, but evolution can not account for them at all.

About 2,000,000 years ago, we are told, man and the monkey family were children of the same parents. These children headed species with an even start. Yet man alone developed personality, consciousness, intelligence, and all the equipment of the soul; all the others remained stationary. This is incredible. It is inconsistent with mathematical probability. Is it likely that one species and one alone out of a million, with similar environments, would reach these high mental and spiritual attainments? No! "God created man in his own image, in the image of God created he him,"-Gen. 1:27. This declaration explains all the difficulties which are insuperable to the evolutionist.

"In the day that God created man, in the likeness of God made he him." This likeness was not a physical likeness as a learned (?) university professor asserted, but a likeness in *knowledge, righteousness and holiness*. No animal is made in the image of God. There is not the trace of a soul in all animal creation. How could the soul of man develop from nothing?

God is still creating new creatures in Christ Jesus, in righteousness and true holiness, which can not come by evolution, for sinful creatures can only grow in sinfulness, until the creative power of God makes them new creatures, as the following study in Eugenics will show: Elizabeth Tuttle, the grandmother of Jonathan Edwards, the eminent scholar and divine, was, according to H. E. Walter, a "woman of great beauty, of tall and commanding appearance, striking carriage, of strong, extreme intellectual vigor, and mental grasp akin to rapacity, but *with an extraordinary deficiency in moral sense. She was divorced from her husband on the ground of adultery and other* IMMORALITIES. The evil trait was in the blood, for one of her sisters murdered her own son, and a brother murdered his own sister, As Richard Edwards, his grandfather, had 5 sons and 1 daughter, by a second wife, but none of their numerous progeny rose above mediocrity, and their descendants gained no abiding reputation, Jonathan Edwards must have owed his remarkable mental qualities largely to his grandmother rather than his grandfather. He was evidently a new creation in Christ Jesus and was cured by grace of all inherited immoralities, so that he became the ancestor of one of the most remarkable families in the history of the world, as follows:—

"Jonathan Edwards was born in 1703. He was strong in character, mentally vigorous and fearlessly loyal to duty. In 1900, of the descendants of Jonathan Edwards, 1394 had been located and the following information in regard to them had been gathered: College presidents, 13; college professors, 65; doctors, 60; clergymen, missionaries, etc., 100; officers in the army and navy, 75; eminent authors and writers, 60; lawyers, over 100; judges, 30; holders of public offices, one being vice-president of the United States, 80; United States senators, 3; managers of railroads, banks, insurance companies, etc., 15; college graduates, 295; several were governors and holders of important state offices."

The claim is also made that "almost if not every department of social progress and of public weal has felt the impulse of this healthy and long-lived family."

"The 'Jukes' family was founded by a shiftless fisherman born in New York in 1720, Since that time the family has numbered 1200 persons. The following facts are quoted from the *records*: Convicted criminals, 130; habitual thieves, 60; murderers, 7; wrecked by diseases of wickedness, 440; immoral women, fully one-half; professional paupers, 310; trades learned by twenty, ten of these learned the trade in prison.

"How much of this expense to the state was due to bad blood we can not say. If the original Jukeses had become Christians we have no doubt that the majority of their descendants would have been humble, but orderly, and possibly useful citizens."

Aaron Burr, a grandson of Jonathan Edwards, lacked but one electoral vote to become president of the U.S. His intellectual standing in Princeton was not equaled by another for 100 years.

Jonathan Edwards was a new creation, as is every other regenerated person.

According to evolution, there can be no new creation. According to the word of God, and the experience of an innumerable host, God is continually creating souls anew, who become "new creatures". Evolution is not in harmony with the Bible nor the experience of the children of God.

Whenever it can be shown that men become more spiritual when they accept the theory, and become more devoted to saving souls as their zeal for the theory increases, the theory will be worthy of more serious consideration. We await the evidence.

Evolution can not account for the spirituality of man, but tends to destroy it where it exists.

41
THE HOPE OF IMMORTALITY

The belief in the immortality of the soul has been well nigh universal, in all ages, and among all nations, and is taught by all religions. Without it, life and death are insolvable mysteries. A doctrine so universal, so well established by reason, ought not to be set aside without the most convincing reasons and the most compelling evidence. Either this universal belief is due to revelation, or the abundance of proof appealing to reason, or both.

A child is born, suffers agonies for weeks and months, and dies. If no future, who can solve the mystery? John Milton writes his immortal "Paradise Lost," and dies. Must his great soul perish? Nero murdered his brother, his sister, his wife and his mother, and multitudes of Christians and lastly himself, and was guilty of a multitude of other shocking crimes; while many of the best men and women this world ever knew suffered persecution and martyrdom for doing good and blessing others. Will they all alike meet the same fate—annihilation—at the hands of a just God?

The immortality of the soul is supported by science. Science teaches the indestructibility of matter. Not all the power that man can bring to bear, can destroy the minutest portion of matter, not a molecule, not an atom, not an electron. The smallest particle of dust visible to the eye contains, we are told, about 8,000,000,000 atoms, and each atom, as complex as a piano,—1740 parts. Not one of these atoms or parts could be annihilated by all the power of a thousand Niagaras.

In all the multiplied chemical changes everywhere in the world, not a single particle, the most worthless, is lost or destroyed. Dissolve a silver dollar in aquafortis, and then precipitate it to the bottom, and not a particle need be lost. If God takes such scrupulous care of the most worthless particle of matter, will he suffer the immortal soul to perish? If he preserves the dust, how much more so the highest of all his creations, the mind that can write an epic, compose an oratorio, or liberate a race. Evolution crushes out of the heart the hope of immortality, and makes man but an improved brute, while Jesus Christ "hath brought life and immortality to light through the gospel."

If evolution be true, when did man become immortal? At what period did he cease to be a brute, and become an immortal soul? Was it before the days of the pithecanthropus, the Piltdown fraud, the Heidelberg man, or the Neanderthal man?

The change was ever so slow and gradual; could the parents, anywhere along the line, be mere brutes and the children immortal human beings? Would it not be impossible to draw the line? Is it not evident that the ape-man could never grow into immortality, or into the image of an infinitely great and glorious God?

If evolutionists *could* give us any convincing evidence that the body of man developed from the brute, they can not prove that the soul grew from nothing to the high mental, moral and spiritual attainments, into the very image of God, and by its own efforts become as immortal as God himself.

After all, did any theory as ridiculously untrue as evolution ever masquerade as science, or ask to be accepted by thoughtful men? Has it as much to support it as the false sciences of alchemy and astrology?

The brute origin of man, infidelity, agnosticism, modernism, atheism and bolshevism, are in harmony, and cooperate in robbing man of heaven and the hope of immortality.

If man believes that he dies as the brute dies, he will soon live as the brute lives, and all that is precious to the heart of man will be forever destroyed. We recoil from such a fate, but live in the serene assurance that such a thing can never be.

42
SIN

Sin is a great fact. It can not be denied. It can not be explained by evolution. It is universal. Every race all nations, with all grades of intellect and culture, civilized or uncivilized, are cursed with sin. All the wrongs, all crimes in the world, all immoralities, are due to sin. Sin causes tremendous destruction of life, property, and character. Why is it universal? When did it originate? Did it originate in all the members of the brute-human race at one time? Did some become sinners, and others remain without sin? Sin must be developed, since brutes have no sin. Why not some of the ape-humans without sin? Does natural selection explain the universal sinfulness of man, on the ground that those who did not have this "improvement" perished? They all died and only sinners were left, hence all survivors are sinners! Sin makes men more fit, and hence sinners only survive! Is evolution simply ridiculous, or a crime?

When in the "ascent of man" did he become a sinner? A million years ago? Judging from the pictures of fierce alleged ape-men, it must have been a long, long time ago. Did all become sinners then? What became of the progeny of those who had not secured the attainment of sin? Why have not other members of the monkey family become sinners? Why do we not hang them for murder? Will they yet attain unto sinfulness?

H. G. Wells, the alleged historian, says, p. 954, Outline of Hist., "If all the animals and man had been evolved in this ascendant manner, then there had been no first parents, no Eden and no Fall. And, if there had been no Fall, then the entire historical fabric of Christianity, the story of the first sin, and the reason for an atonement upon which the current teaching based Christian emotion and morality, collapses like a house of cards."

Evolution claims that man fell up and not down. It denies almost every truth of religion and the Bible, as well as of experience. "Man is falling upward, he is his own Savior, he is ever progressing, and has no need of a Savior." Contrast this with the sublime statements of the word of God concerning the creation and the fall of man.

Evolution is charged with explaining all phenomena pertaining to man,—soul and body. It exhausts itself in trying to show that the body of man may *possibly* be developed from the brute. It fails miserably. The problem of accounting for the soul of man with all its equipment is so much more difficult, that little or no effort is made to account for it, virtually confessing that the much-exploited theory of evolution can not possibly be true, when applied to the soul as well as the body.

43
REDEMPTION

Evolution does not account for sin. Much less does it have any cure for sin. If sin marks progress or advancement, of course, its cure would be retrogression. But how can sin be cured? What answer has evolution? Culture, education, refinement, favorable environment. These are all desirable, but no cure for sin. Some of the most cultured, educated and refined, were the greatest monsters that ever lived. Wholesale murderers like Nero, Alexander and Napoleon, had a good degree of education and culture. Nathan Leopold and Richard Loeb, who murdered Robert Franks in Chicago, were among the most brilliant graduates of universities. Friends say they were led on to atheism and crime by the reading of modernist books. No doubt, the doctrine of evolution, taught so zealously in the universities, played a large part.

Human efforts and human devices have utterly failed to cure sin. The human will is too feeble to resist its power.

The Bible, which evolution undermines, teaches us there is a cure for sin. The divine Son of God saves us from our sins, cleanses and purifies our natures, and fits us for happiness and service in both worlds. Jesus offers the only practical plan of salvation from sin. The Bible plan of redemption is the only plan that works.

Paul, a murderer, with his heart full of malignant hate, and his hands stained with blood, greedy to imprison men and women, "breathing out threatening and slaughter," looks to Jesus by simple faith, and is changed into a gentle and loving Christian, rejoicing in suffering and persecution. He rose to such heights, by the help of Jesus, that he loved his enemies, and was willing to be damned, if that would save their souls. What glorious men the apostles became by the transforming power of Christ! What grand men and women the long line of martyrs were. The men and women who have blest the world most, have been believers in the Bible, and not in evolution. Perhaps a million martyrs have died for Christ. Where are the martyrs for evolution?

Augustine was redeemed from a life of vice and dissipation, blessed the world with his writings, and became one of the greatest leaders of thought in all ages. John Bunyan was so profane that the most vicious would cross the street to avoid him. The gospel made him one of the holiest of men. His Pilgrim's Progress has been translated into hundreds of languages, and read by millions. John G. Woolley was a maudlin drunkard, intent on taking his own life,—friends, money, character, and reputation lost,—but was converted and preached, with burning eloquence, the gospel of temperance and prohibition around the world.

Elijah P. Brown, a zealous infidel, heard Mr. Moody preach on the love of God, found the Savior, and became a brilliant defender of the faith.

Chundra Lela, the daughter of a Brahman priest, spent a fortune and lived a life of self inflicted torture, seeking salvation at all the great shrines of India, but found none, until she heard the simple story of Jesus from the lips of a missionary. That matchless name gave her victory over sin, and transformed her into a saint and soul-winner for Christ. Maurice Ruben, a successful Jewish merchant of Pittsburgh, rejected Christianity and the Jewish religion as well. He was converted, ostracised, persecuted, thrust into an insane asylum unjustly, and told he must give up Christ or his wife and child. He chose Christ. His family soon became Christians and joined him in the great Jewish mission in Pittsburgh.

In a single night, the mountain floods in India caused the death of the six children of Rev. D.H. Lee,—only one living a short time to tell the story. They were all musicians. Out of the awful silence of that home, Mrs. Lee sent to American papers, a triumphant pean of praise to God. She was sustained by the power of God, so that she could kiss, in loving devotion, the hand that smote her. The Lee Memorial Orphanage, of Calcutta, stands as their monument.

Holy Ann, of Canada, was so profane and such a terror, that this name was given her in derision. Touched by Christ, she became so sweet a saint, that all regarded her as holy indeed.

Geo. Long, a denizen of the underworld, a victim of strong drink, cocaine, opium and morphine, ruined in body and soul, was redeemed and freed from these desperate vices, and made a successful soul-winner for Christ.

These are a few of that "multitude that no man can number" who have been delivered from the power of sin, and have overcome by faith in Jesus.

If evolution be true it should be no hindrance but a great help. How many drunkards have been saved by a belief in evolution, and how many have been greater soul winners by such belief? How many criminals have been saved by acceptance of the theory? Many have been made criminals, unbelievers, infidels, agnostics and atheists by it; how many have been made Christians? Can any one be named who has been made a more earnest and successful soul winner, or a sweeter saint, by espousal of the doctrine? If one blank page were set aside for a list of all victims of sin and vice and crime, who were redeemed by faith in evolution, the space would be wasted. Is there any comfort in it to the dying, any help to the living? Would any evolutionist preacher read to the dying, the so-called classic passage from Darwin, showing that every living thing on the tangled bank came from one germ without any assistance from God? Is there any choice passage in all their books, fit to be read to the dying, or to a man in trouble, or in need of salvation? Is there anything to put hope in the breast, or inspire a man to a holy life? Anything to lift up a man sodden with sin, and redeem him from the fetters that bind him?

To give up the tested power of the gospel and to accept instead, the worthless guesses of evolution, ruinous in life and powerless in death, would be a sorry exchange indeed.

44

EV. AIDS INFIDELITY AND ATHEISM

Many evolutionists frankly declare that the purpose of evolution is to destroy belief in God, or his active control of his creation. Prof. H. F. Osborn, of N. Y., a leading evolutionist, says, "In truth, from the period of the earlier stages of Greek thought, *man has been eager to discover some natural cause of evolution, and to abandon the idea of supernatural intervention in the order of nature.*" Other evolutionists openly announce their antagonism to the Bible and Christianity. Clarence Darrow, in the Tenn. trial, called Christianity a "fool religion."

Darwinism has been declared an attempt to eliminate God and all evidence of design and to substitute the old heathen doctrine of chance. With this announced purpose in view, we are not surprised to learn from Prof. J. H. Leuba that one-half the professors teaching it did not believe in God nor the immortality of the soul; and that there is a rapid increase in the number of students who have discarded Christianity as they progress in their course,—Freshmen, 15%; Juniors, 30%; Seniors, 40 to 45%. Children of Christian homes, taught to believe in God and Jesus Christ, are led into infidelity and atheism rapidly, as they progress in their course. It makes one shudder to think what the future will be, if atheism and infidelity are taught in the guise of science. And the statistics show that evolution is one of the most fruitful sources of unbelief. What the students are taught today, the world will believe tomorrow. How great the havoc caused by a comparatively few infidel or atheistic professors!

Dr. C. W. Elliott, a Unitarian, announced with apparently great glee, that already the young men and young women do not believe the story of the creation of Adam and Eve. The leaders of Bolshevist Russia said to Dr. Sherwood Eddy, with brutal frankness, "The Communist party, the only party allowed in Russia, is 100% atheistic. If a man believes in God, he can not be a member of the party." Russia is an example of a country where atheism is taught in the public schools, and we are moving all too fast in the same direction. The Red Army shot to death 500,000 men in Russia. The horrors of the French Revolution may be outdone, if we do not awake to our danger. Russia is cursed with a doctrine offensive alike to the Christian, the

Jew, the Mohammedan and even the deist. In America the same condition may be brought about, more stealthily and more effectually in the name of science. Indeed, the Russian atheists feel the necessity of adopting the American method as more effective. An Associated Press dispatch of Dec. 24, 1924, states that Zinovieff, a Soviet leader, admitted that the Communists had gone too far in their efforts to establish atheism by *force*, but he adds, *"We shall pursue our attacks on Almighty God in due time, and in an appropriate manner. We are confident we shall subdue him in his empyrean. We shall fight him wherever he hides himself....* I have been informed that not only young Communists, but Boy Scouts, are mocking people who are religious. I have also been told that *groups of Boy Scouts have even imprisoned whole congregations in church while they were worshipping*! Our campaign against God and religion must be carried out in a pedagogic way, not by violence or force." Do we want such a situation in America? We are drifting that way.

Evolution has no quarrel with atheism, agnosticism, modernism, or any other species of infidelity. *Its quarrel is with Christianity and the Bible.* Why should we wish to harmonize Christianity with evolution, when the theory can not possibly be true? Prof. Newman says, "Readings in Evolution," p. 8, "Contrary to a widespread idea, evolution (in what sense?) is by no means incompatible with religion (Christianity?).... The majority of thoughtful theologians (whew!) of all creeds are in accord with the evolution idea."

Dr. W. W. Keen says, "I believe in God and evolution." An infidel, a deist, even a heathen can say that. To harmonize evolution with Christianity is quite a different problem. Prof. Coulter, of Chicago University, endeavors to show where "religion and evolution meet." But the "religion" is the religion of the infidel, not of the Christian. How can a theory which denies the creation of Adam and Eve and any intervention and control by the Creator, be harmonized with Christianity?

Rev. F.E. Clark, President of the World C.E., says, "The Darwinian theory, whatever it may be called today, has doubtless unsettled many minds. A hazy agnosticism has often taken the place of strenuous belief." He is in a position to know.

A beloved friend, president of a prominent college, an evolutionist and a modernist, in a letter to the writer, claimed that evolution is nearest the truth, and those who believe it are nearest to "Him who is the Way, the Truth and the Life." If this is true, how many evolutionists are more spiritual, more earnest, and more successful on that account, in winning souls to Christ?

No doubt many have been made infidels and atheists. How many souls have been won to Christ by Osborn, Newman, Conklin, Darrow, Lull,

Shull, Scott, Coulter, Metcalf, Kellogg, Nutting, Thompson, Castle, Chapin, and all other prominent evolutionists? If evolution is nearest the truth, the number of their converts to Christ should be greatly increased. We await the information, which we do not have at hand, to see if the contention of our friend is correct.

Mrs. Aimee Semple McPherson preaches daily in the Angelus Temple, Los Angeles, Cal., which seats 5300 people. Often standing room is at a premium. Many souls are saved (over 14,000 in 1924), and thousands are healed in answer to prayer. What a tremendous loss to humanity, if the gospel of Christ had not saved her from the infidelity and atheism of evolution! She writes as follows of her conversion: "The writer went to one of the services being held in my home town, by the Irish evangelist, Robert Semple, and entered the meeting *practically an infidel, having studied Darwinism, atheistic theories until faith in God's word was shaken.* Never will those moments be forgotten. One could feel the power of God, the moment one entered the building. Such singing, hands uplifted, faces radiant, such Amens and Hallelujahs, such power and fervor back of every word that was spoken, such exaltation of the deity of Christ, the necessity and power of the atoning blood, the second Coming of Christ, the power of the Holy Spirit to energize and get the believer ready for his coming, gripped and stirred the heart.... Never, never, can the writer forget that hallowed hour, when, kneeling by a Morris chair in the home of a friend, early in the morning, with uplifted arms, she prayed and felt for the first time, the tremendous inflowing power of the Holy Ghost." Behold, the power of evolution to ruin, and of Christ to save!

Evolutionists are, as a rule, modernists; and modernists are evolutionists, and are reckless in their zeal to destroy the faith of the young committed to their care. We select the following 3 illustrations from a single article in the PRESBYTERIAN:

1. "A father sat *in this office*, a minister above middle life, his eyes full of tears, and his soul full of groans, as he told how he had sent his son, who had been an orderly Christian boy, to a supposedly Christian college. When the boy returned home, after graduation, he informed his father that through instruction received, he had lost his faith, and believed none of those things he had been taught at home. The father was so shocked and overcome he could make no reply, but asked his son to kneel and pray with him as they used to do. The son refused, and said he no longer believed in prayer."

2. "A good Christian father desired to give his young daughter the best educational advantages. She planned to be a missionary. He sent her to a

well-known college, considered Christian. This college had a Bible chair, but of the destructive, critical type. The young student absorbed what she was taught. She lost all reverence for the Bible and rejected it. She entirely lost her faith which she had learned from her father and mother. She gave up her mission plans, and developed into a Socialist. When about to graduate, she wrote her father frankly, that she had given up the faith he had taught her, and she was going to live with a man without marriage, as she did not believe in marriage; The father visited her and protested. She smiled and called him an old fogy. She only consented to marriage when threatened with the civil law."

3. "Another case reported to us by another father:—His son, attending a so-called Christian college, reported that one of the professors declared that they and himself were hypocrites, because they attended chapel every morning where they were told that if they believed and did such things, they would some day go to another world and play on a harp. But if they did not, they would burn. This he declared was all bosh. Then he called attention to the teachings in the college, that *man in his body developed from a lower animal, but that man had no soul.*"

Yet some colleges and universities ask Christian people to give large sums, with no guarantee that evolution, infidelity and atheism will not be taught. Is it any wonder that Christian parents tremble while their sons and daughters run the gauntlet of infidel professors?

45

EV. WARS WITH CHRISTIANITY

Evolution leads to infidelity and atheism, and is therefore a foe to Christianity. It denies the doctrine of special creation, and opposes the religion of the Christian, the Jew and the Mohammedan. Why should not all these religions unite against the false and unsupported theory that would make havoc of them all?

If evolution could be shown reconcilable with Christianity it would be lifted into respectability, but what would be the gain to Christianity? The Christian religion is reconcilable with all true science, and hails every true science with joy. The church loves true science, but hates a lie that poses as the truth. Christianity is readily reconcilable with the true sciences of Astronomy and Chemistry, but we do not try to reconcile it with the corresponding false sciences of astrology and alchemy. Why should we be concerned about such a reconciliation, since all the evidence offered in favor of evolution is not worthy of serious consideration? The facts hotly contest every guess. There is no conflict between Christianity and science. But evolution is not science. It is not knowledge. It is not truth. It is not proved. It is not certain. It is not probable. It is not possible. How can the serious student escape the conviction that evolution has not one chance out of a thousand, or even out of a million, to be a possible theory, and none whatever to be a probable or proven theory? It offers not one convincing argument. The evidence against the theory shows that it has not yet been proven and never can be.

The present population of the globe shows the unity of man in the days of Noah, and that the human race could not have begun 2,000,000 years ago, nor 1,000,000, nor 100,000, nor even 10,000. And no evidence that the evolutionist can bring to bear *now or hereafter* can ever set aside this mathematical demonstration. This one argument is sufficient to shatter evolution, if there were no more. But the whole fifty arguments in this book rush to the support of this one. They all harmonize with the Bible statements, but not one of them with the false and baneful theory of evolution. And no erroneous guess that they can make will escape mathematical detection. Why should we gratify the clamor of evolutionists, and seek to reconcile

Christianity with a theory so manifestly false? To be worthy of acceptance, it must satisfactorily answer every one of the fifty arguments in this book and many more. Can it do so?

Evolution carried to a logical conclusion would destroy every thing precious to the heart of a Christian. It denies the real inspiration of the Bible. It makes Moses a liar. It denies the story of creation, and substitutes an impossible guess. It denies miracles, the providence of God, the creation of man and beast, and God's government and control of the world. It laughs at the Virgin Birth and makes Christ a descendant of the brute on both sides. It denies his deity, his miracles, his resurrection from the dead. It joins hands with agnosticism, modernism, and other forms of infidelity and atheism and gives them the strongest support they have ever had. All these hail evolution's advent with exceeding great joy. It has the closest affinity with the wildest and worst theories ever proposed.

Its writers and proponents turn infidel and atheist. Its teachers and advocates lose their belief in God and the immortality of the soul. The young men and women who are taught, abandon the faith of their fathers and join the forces of unbelief. To be sure, some are saved by inconsistency, and still maintain their faith, but the havoc is great. It would strip Christ of his Deity, reduce him to the dimensions of a man, and make his religion powerless to save. The men who tore the seamless coat from the dying Christ did a praiseworthy act, in comparison to those who would strip him of his deity and glory, for these are the garments of God!

The ruffians at the foot of the cross gambled for a mere human garment, but there are evolutionists who would "trample under foot the blood of the Son of God, and count it an unholy thing." Those who would rob the world's redeemer of his power and divinity, while speaking patronizingly in praise of his human traits, do but insult him with the vilest slander, which makes the derision of Calvary seem like praise.

We were not surprised to learn that, in the Tenn. trial, evolution was defended by agnostics, who made their chief attack on the Bible and revealed religion; and the school, the home and religion were defended by men of high Christian character. Had Mr. Darrow as earnestly defended Christianity and Mr. Bryan as earnestly opposed it, millions would have held up their hands in astonishment. But the alignment was natural, and opened the eyes of multitudes to the fact that evolution is a friend to infidelity and a foe to Christianity. Their objection to prayer during the sessions of the Court shows that they hated what God loves.

Christianity withstood ten fiery persecutions, lasting 300 years, at the hands of the Roman Empire, the mistress of the world. The church was purified, and grew and multiplied. Numerous heresies arose but all yielded to the truth. Sin and corruption, formality and worldliness, failed to hinder the triumphant march of the church of God.

Infidelity made a fierce attack in the eighteenth century in its own name, and lost. But the most dangerous attack ever made is on, by evolution claiming the name of science and modernism claiming the name of religion. This f. a. d. is truly for a day. God will win. Truth will live and error will die. But too many precious souls will be lost unless the world awakes to see its danger soon.

Mr. Bryan, in his last message, said: "Christ has made of death a narrow starlit strip between the companionship of yesterday and the reunion of tomorrow. Evolution strikes out the stars, and deepens the gloom that enshrouds the tomb.".... "Do these evolutionists stop to think of the crime they commit when they take faith out of the hearts of men and women and lead them out into a starless night?"

Evolution wars with the religion of the Jews also. It attacks the Old Testament, dear alike to Christian and Jew. The Jews were the chosen people of God, and have played a large part in the history of the world. We gladly clasp hands with them against the common foe. David speaks for Jews and Christians in the 8th Psalm. In contrast to evolution, which degrades man to the level of the brute, he declares that man is but a little lower than God, (Heb. Elohim). The revisers had the courage so to translate it. David under inspiration wrote better than he knew, and in absolute harmony with modern science:

"When I consider thy heavens, the work of thy fingers, the moon and the stars which thou hast ordained, what is man (how great must he be) that thou are mindful of him (among thy great and marvelous works)? And the son of man that thou are a companion to him? For thou hast made him but little lower than God, and crownest him with glory and honor. Thou madest him to have dominion over the works of thy hands; thou hast put all things under his feet; all sheep and oxen, yea, and the beasts of the field; the fowl of the air, and the fish of the sea, whatsoever passeth through the paths of the seas." All animals confess the dominion of man since the strongest and fiercest flee from his face. Who would prefer the "string of stuff" that would place man below the brute, to the lofty description of the Hebrew Psalmist placing him a little lower than God?

Hon. William J. Bryan, when attending the Presbyterian General Assembly in Columbus, Ohio, in 1925, enclosed, in a letter to the writer, a copy of his address in John Wanamaker's Church, Philadelphia, on evolution and modernism, from which we select the following:

"All the modernists are evolutionists and their hypothesis of creation gives man a brute ancestry and makes him the apex of a gradual development extending over millions of years. This hypothesis contains no place for, and has no need of, a plan of salvation. It is only a step from this philosophy to the philosophy of the atheist who considers man 'a bundle of tendencies inherited from the lower animals,' and regards sin as nothing more serious than a disease that should be treated rather than punished. One of the gravest objections to the doctrine of the modernists is that it ignores sin in the sense in which the Bible describes sin. Modernists ignore the cause of sin, the effects of sin, and the remedy for sin. They worship the intellect and overlook the heart, 'out of which are the issues of life.' No evangelical church has ever endorsed a single doctrine of the modernists.

"Evolution is the basis of modernism. Carried to its logical conclusion, it annihilates revealed religion. It made an avowed agnostic of Darwin (see in his 'Life and Letters' a letter written on this subject just before his death); it has made agnostics of millions and atheists of hundreds of thousands, yet Christian taxpayers, not awake to its benumbing influence, allow Darwinism to be injected into the minds of immature students, many of whom return from college with their spiritual enthusiasm chilled if not destroyed.

"When we protest against the teaching of this tommy-rot by instructors paid by taxation, they accuse us of stifling conscience and interfering with free speech. Not at all; let the atheist think what he pleases and say what he thinks to those who are willing to listen to him, but he cannot rightly demand pay from the taxpayers for teaching their children what they do not want taught. The hand that writes the pay check rules the school. As long as Christians must build Christian colleges in which to teach Christianity, atheists should be required to build their own colleges if they desire to teach atheism.

"With from one to three millions of distinct species in the animal and vegetable world, not a single species has been traced to another. Until species in the animal and vegetable world can be linked together, why should we assume without proof that man is a blood relative of any lower form of life? Those who become obsessed with the idea that they have brute blood in their veins devote their time to searching for missing links in the hope of connecting man with life below him; why do they prefer a jungle ancestry to creation by the Almighty for a purpose and according to a divine plan? Why will they travel around the world to find a part of a skull or remnants of a skeleton when they will not cross the street to save a soul?

"How can intelligent men and women underestimate the Christ? He is no longer a wandering Jew with a few followers; He is the great fact of history and the growing figure of all time—there is no other growing figure in all the world today. Men—the greatest of them—rise and reign and pass away; only CHRIST reigns and remains. They shall not take away our Lord. The Christian Church will not permit the degrading of its founder; it will defend at all times, everywhere and in every way, the historical Christ. It believes that 'there is none other name under heaven given among men, whereby we must be saved.' No diminutive Messiah can meet the religious need of the world today and throughout the centuries. Christ for all and forever, is the slogan of the church. There has been apostasy in every age; attacks upon Christianity have been disguised under cloaks of many kinds, but it has withstood them all—'The hammers are shattered but the anvil remains.' The church will not yield now; it will continue its defense of the Bible, the Bible's God and the Bible's Christ until 'every knee shall bow and every tongue confess.'

"While it resists the attacks upon the integrity of God's Word and the divinity of the Saviour, it will pray that those who are now making the attack may come under the influence of, and yield their hearts to, Him whose call is to all, whose hand is all power and who promises to be with His people 'always, even unto the end of the world,' The Apostles' Creed which has expressed the faith of the Christian Church for so many centuries shall not be emasculated by modernism.

> "'Faith of our fathers! living still
> In spite of dungeon, fire and sword;
> O how our hearts beat high with joy
>> Whene'er we hear that glorious word—
> Faith of our fathers! holy faith,
> We will be true to thee till death'!"

46
CAMOUFLAGE OF TERMS

During the late world war, objects were concealed and the enemy deceived, by "camouflage." Many undertake to deceive or to hide their meaning by a camouflage of terms. These terms are chosen to conceal or deceive. Terms that suggest advance, improvement, learning, science, etc., are used to describe unworthy theories, beliefs and movements. It is an unfair trick to win and often meets with undeserved success.

Evolution in the sense of growth and development, is true of a part of animal and plant life, and in this sense is undisputed. Some speak of the growth of a child and of all progress, as evolution. In the sense at issue, it means the development of all the 3,000,000 species of animals and plants, from one or a few primordial germs, without design or intelligence, or the aid of a Creator. A distinguished surgeon declares that evolution from the monkey is mere non-sense but that life is a constant evolution,—two senses in the same sentence. Such confusion of meaning brings science into disrepute. The meaning is shifted to suit.

Science means knowledge. We are glibly told that science teaches the evolution of man when it teaches nothing of the kind. A mere theory is not science until proven. A man does not become a scientist by advocating an unproven theory, but by making some notable contribution to knowledge. These self-appointed scientists recklessly declare that the "consensus" of science favors evolution. We oppose evolution not because it *is* science, but because it is *not* science. There is no conflict between Christianity and real science, but a fight to the death with "science falsely so called."

Religion is often taken to mean deism, or infidelity as well as Christianity. They show us "where evolution and religion meet," provided deism or infidelity is religion, but not, if Christianity is religion,—an inexcusable confusion of terms.

Law is sometimes spoken of as if it had intelligence and power. Sometimes as a subordinate deity, or agent of God, or an indefinite principle. Darwin says:—"Plants and animals have all been produced by laws (?) acting around us." That is impossible, since "laws" can produce nothing. He evidently gives to laws the credit that belongs to God.

Nature, in like manner, is often used as a substitute for God, to avoid the mention of His name.

Modernism is a fine sounding word, suggestive of learning and culture and the last word in science, but doubts or denies many of the essential doctrines of the Christian religion. It is infidelity pure and simple and of the most dangerous kind, camouflaged under this attractive name. Who can deny the statement that the only thing modern about modernism is its hypocrisy? It is ancient infidelity pretending to be a Christian view. Bearing the Christian flag, it attacks Christianity. Modernists are evidently ashamed of a name which fitly describes their views, and seek another. Infidels have tried to win under their own name. They have failed. Will they succeed under the camouflaged name of modernism? Camouflaged under an attractive name, modernists doubt or deny the real inspiration of the Bible, the Virgin birth of Jesus, his deity, his miracles, his bodily resurrection, the resurrection of the dead, and his personal second coming to judge the quick and the dead. Some modernists reject a part of these great truths, and some reject all.

Liberal is another term stolen by infidels ashamed of their own name. They are no more liberal in a good sense than others.

A Rationalist is not entitled to the term, because he is often more innocent of reasoning than his opponents. Reason is not opposed to revelation. We believe in an inspired revelation, because it is reasonable to do so. Rationalism is another camouflage for infidelity. We can have some respect for an honest professed skeptic, but how can we respect a man who insists on adding hypocrisy to his infidelity, that, by so doing, he may make greater havoc of the church? Modernists give such a diluted interpretation to inspiration, to the statements of Scripture, and the Apostles' Creed, and the creeds of the churches, that all may mean little or nothing, and the floodgates of infidelity and atheism are opened wide.

It has been truly said, "If the Bible is not really inspired, it is the greatest fraud ever perpetrated on mankind; for, from lid to lid, it claims to be the word of God." Likewise, if Moses was not inspired, he was the greatest liar of history.

Every variety of infidel and species of atheist will rejoice, if evolution be accepted,-whether modernists, liberals, rationalists, or simple unbelievers on their way to the bottomless pit. If evolution wins, Christianity loses and the church fails.

We hope that scientists will consign to innocuous desuetude their camouflaged sesquipedalian vocabularies, and tell us what they mean in short words, so we all may know what they say.

47
WHAT ARE WE TO BELIEVE?

Some would have us believe there is no God; or that matter is eternal; or that matter was evolved out of nothing; or that all things came by chance; or that there is nothing but matter,—no God, no spirit, no mind, no soul.

Some would have us believe that God created nebulous matter, and then ceased to control the universe; that life developed spontaneously; that species developed by chance, or natural selection, or by a powerless "law," from one primordial germ. Others say that all the countless exhibitions of design by a matchless Intelligence, are to be explained by a causo-mechanical theory, which means the theory of blind unintelligent chance, without purpose or design or interference of God. Some say that God may have created one germ or at most 4 or 5, and that 3,000,000 species of plants and animals developed from this microscopic beginning. We are asked to believe that some plants became animals, or some animals became plants, or that all plants and animals came from the one germ they allowed God to create. They say that all species developed by growth, but do not explain why we still have the one-celled amoeba, the microscopic bacilli of plant life, and the microscopic species of animal life. Many geologic species are largest at the beginning; many ancient animals were much larger than their successors; and the reptilian age was noted for animals of enormous size. Yet they want us to believe that growth is universal.

They ask us to believe, without proof, that some marine animals evoluted into amphibians, some amphibians became reptiles, some reptiles developed hair and became mammals, and some reptiles developed feathers and wings and became birds; some mammals became monkeys, and some monkeys became men. For evidence of this, there is not a single connecting link to show the transformation. Geology furnishes no fossils of the millions and billions of connecting links that must have existed. For the scheme would require not only millions of links between man and the monkey, but also millions between each of the 8 great changes from matter to man. Yet we are asked to accept these fantastic and impossible speculations as "science," though it lead to infidelity and atheism and bolshevism and anarchy and chaos, wreck religion, make havoc of the church, and send countless souls to the lost world. What wonder that the soul recoils with horror from such an atheistic theory.

48
WHAT CAN WE DO?

Evolution, leading to infidelity and atheism, is taught in many universities, colleges and high schools, and even in the lower grades of the public schools. It is taught also in some theological seminaries. It is proclaimed in some pulpits. Some of its devotees, who have slipped into places of power and influence, urge it with a zeal worthy of a better cause. The public libraries are crammed with books teaching it, with few, if any, opposed. Strange to say, it is advocated by some religious newspapers, along with modernism and other varieties of infidelity. Some secular newspapers seem eager to publish, on the front page, attacks on orthodoxy, and articles favoring the wildest claims of evolution. They call evolution science! What are we going to do about it? Shall we supinely submit, or do all in our power to oppose, check and suppress so pernicious a theory? What can we do?

We can refuse to patronize or endow such institutions as teach this or other forms of infidelity and atheism. We can aid those only that are safe. Much money that was given by devout Christians to colleges and seminaries, has been prostituted to teach what the donors hated, and to do great harm. The faculty and trustees can do much to eliminate false teaching, if they will. Use all possible pressure to bring this about.

Evolution is taught in many high schools supported by the taxpayers' money. This should not be tolerated. Text books declare that man is descended from the brute, as if there were no doubt about it! Laws should be enacted and courts appealed to, to protect the youth. The recent decision of the Supreme Court of the United States in the Oregon case, gives strong hope that the teaching of evolution would not be permitted, if a case were carried up to the highest court. It should be done. If Christianity cannot be taught in the public schools, must we submit to the teaching of infidelity and atheism in the name of science? Intolerable outrage! In New York 15,000 people, on a recent Sunday, shouted for atheistic bolshevism, and condemned the United States government. A theory that encourages such a belief should not be taught. When the people awake to see the baneful effects, they will smite the fraud to the earth. Protests should be made to Boards of Education, superintendents, and all in authority. The power of public opinion should be brought to bear. Two states already have forbidden such instruction, and others will, no doubt, follow. The Associated Press, in this morning's papers, calls the struggle a contest between religion and science, and thousands of shallow thinkers will believe that evolution is really science!

We quote from Mauro's "Evolution at the Bar," p. 71: "A parent writing to a religious periodical, tells of a text book brought home by his seven-year-old boy, the title of which was, 'Home Geography for Primary Grades.' Discussing the subject of birds, this text book for primary grades says: 'Ever so long ago, their grandfathers were not birds at all. Then they could not fly, for they had neither wings nor feathers. These grandfathers of our birds had four legs, a long tail, and jaws with teeth. After a time feathers grew on their bodies, and their front legs were changed for flying. These were strange looking creatures. There are none living like them now.'" Would any one who would teach a little child, the extremely improbable story that reptiles became birds, hesitate to teach that monkeys became men and that the story of creation was false?

Much can be done by the church authorities in refusing to license or ordain men who believe in any species of infidelity, or who have attended heretical seminaries. They should give their consent for candidates to attend only colleges, universities or seminaries that can be trusted. Congregations should know, before they call a pastor, that he is orthodox. Ministers are to preach the Gospel not infidelity.

Taboo all heretical religious papers; support those that defend the truth. Let infidels maintain infidel papers and build infidel colleges. Not one dollar to propagate infidelity! Make your one short consecrated life count for truth and righteousness. Many Christians are guilty of the great sin of indifference. In this greatest of all contests in which the Church was ever engaged, no one should be a slacker.

Many public libraries have 20 to 50 books in favor of evolution, and but one or two, if any, opposed. If dangerous books, like Wells' "Outline of History", McCabe's "A. B. C. of Evolution", and the works of Darwin, who doubted his own theory, and of Romanes, who renounced evolution and embraced Christ, can not be eliminated, libraries, in all fairness and in the interest of truth, should have an equal number in reply. Insist that librarians get a copy of this book, and other anti-evolution books, especially those mentioned herein; also other good books.

The author and publisher of this book will give 50% commission for selling it, and will mail two copies for $1.00 to all who will become agents. If you can't be an agent, you will do great good by securing another. A copy should be in the hands of every student, so he can discuss evolution with his teacher; and in the hands of every teacher, lawyer, doctor, minister, lawmaker or other professional man, of every parent whose children are liable to be taught the dangerous doctrine. It will be useful in removing error and in promoting the truth. Agents should canvass every school, college, university, seminary; every convention, conference; every religious and educational gathering. A copy should be in every library.

Every dollar of profit from the sale of this book will be given to Missions, to be loaned perpetually to help build churches, and to preach the Gospel in the secular newspapers of the world, and to distribute this book free. Every $1000 so loaned to churches at 5% compound interest, in 300 years, will, together with the accrued interest, aid in building 8,229,024 churches, by a loan of $1000 each for 5 years, and the new principal at the end of 300 years will be $2,273,528,000.

After four struggles, the writer was led to give the one-tenth, then the unpaid or "stolen" tenth (Mai. 3:8), then to consecrate the nine-tenths, and, lastly, to give all above an economical living. Many another consecrated Christian, on fire for God and burning with fury against all forms of infidelity, can do incalculable good by sending this book free to as many libraries, students, teachers, ministers, lawyers and doctors as possible. For this purpose, the publisher will mail the book to large numbers, for 20c each; your $1 sends a $1 book to 5. For $2000, for example, a copy will be mailed to the 10,000 ministers of the Presbyterian church, U.S.A.; for $4,000, to the 20,000 pastors of the Methodist Episcopal church; for $1000, a copy to 5000 public libraries in the United States and elsewhere; or to 5000 students, teachers, ministers, lawyers, doctors, lawmakers, etc. Smaller sums in proportion. What great good a heroic giver, in every land, could do with $1000 or $10,000 or $100,000! With 1,000,000 copies, we would wake the world!

A Canadian farmer gives $1000 to mail one to 5000 Canadian ministers and libraries. Who will give $2,000 to send one to 10,000 lawmakers in U.S.?

—Ministers, students, teachers, parents, yes, ALL are urged to be agents, employ sub-agents, earn wages, and do good. To agents, booksellers, libraries, churches, S.S.'s, organizations and societies needing funds, 2 to 25 mailed to any land, for 50c each cash; 25 or more, 40c—60% profit; 100 or more, 30c—70% profit! Books are the best outfit,—try 25 (show p. 76). To periodicals (for sale or premium), 30c. Special terms to general or national agents, speakers, publishers, colleges, seminaries, etc. Editors are hereby given permission FREE to use any selections. Add to each: From 'EVOLUTION DISPROVED' (cloth $1) by per. the author and pub., Rev. W. A. Williams, Camden, N.J. *Mail marked selections and reviews.*

The fight is on. Only about 2% of the members of evangelical churches, it is said, are modernists and evolutionists. Let the rest assert their rights and say: "Common honesty requires you to restore to orthodoxy the institutions you have purloined. We demand them back. Henceforth you shall not steal our colleges, seminaries and public schools, and make our children infidels and atheists. You shall not, with our consent, capture our pulpits, and strip the world's Redeemer of his power and glory."

49
PROBLEMS FOR REVIEW

The following problems, when solved by the reader, will deepen the conviction that evolution is impossible. The erroneous guesses by evolutionists may be checked up and disproved by mathematical problems. No stronger proof could well be devised. For pattern solutions, refer to the preceding text. A reward will be given to the first person who points out a material error. Test, verify or correct the following solutions:—

1. If the first human pair lived 2,000,000 years ago, as the evolutionists claim, and the population has doubled itself in every 1612.51 years (one-tenth the Jewish rate of net increase), what would be the present population of the globe? Ans. 18,932,139,737,991 followed by 360 figures; or 18,932,139,737,991 decillion, decillion, decillion, decillion, decillion, decillion, decillion, decillion, decillion, decillion; or 18,932,139,737,991 vigintillion, vigintillion, vigintillion, vigintillion, vigintillion, vigintillion.

2. If the first human pair lived 100,000 years ago (a period much less than evolution required), what Would be the present population at the same low rate of increase? Ans. 4,660,210,253,138,204,000; or 2,527,570,733 times as many as are living now.

3. At the above rate of increase, how many human beings would have survived in the 5177 years since Noah? Ans. 9. How many Jews, in the 3850 years since Jacob's marriage? Ans. 5.

4. If the human race doubled its numbers every 168.3 years since Noah became a father (5177 years) what would be the population of the globe? Ans. 1,804,187,000,—just what it is.

5. If the Jews doubled their numbers every 161.251 years since Jacob's marriage (3850 years ago), how many Jews would there have been in 1922? Ans. 15,393,815, just the number reported.

6. What guess of man's age can stand the test of mathematics? Ans. Not a single guess ever made assigning a great age to man,—nothing greater than the age indicated by the Scriptures; 2,000,000, or 1,000,000, or 100,000 years are clearly out of the question.

7. If life began 60,000,000 years ago, and the human race 2,000,000 years ago, how much sub-normal should have been the brain and mind of man at that time? Ans. 1/30 or 3-1/3%; or 96-2/3% normal; or 1450 c.c., counting 1500 c.c. normal,—more nearly normal than many nations now.

8. How much if life began 500,000,000 years ago? Ans. .4%; or 99.6% normal; or 1494 c.c., far more c.c. than a large part of mankind can claim.

9. If man had, in 58,000,000 years, developed only the same skull capacity as the other members of the simian family (not over 600 c.c.), how much must he have gained in 2,000,000 years? Ans. 900 c.c., which is a development 43.5 times as rapid in 2,000,000 years as in the 58,000,000 years preceding. How could that be?

10. If life began 500,000,000 years ago, how would the rapidity of skull and brain development in 2,000,000 years compare with that of the 498,000,000 years preceding? Ans. 373.5 times as great.

11. If the skull of the pithecanthropus was two-thirds normal, or 1000 c.c., how many years ago must it have lived, in case life began 60,000,000 years ago? Ans. 20,000,000; in case life began 500,000,000 years ago? Ans. 166,666,666.

12. If the Piltdown "man" had a normal skull capacity of 1070 c.c., as claimed, how long ago did he live, if life began 60,000,000 years ago? Ans. 17,200,000 years. If 500,000,000 years ago? Ans. 143,333,333 years.

13. If the Neanderthal man had a capacity of 1408 c.c. (assigned by Dr. Osborn), how many years ago must he have lived if 60,000,000 years have passed since life began? Ans. 3,680,000; if 500,000,000 years? Ans. 30,666,666. If 1800 c.c. be taken as normal instead of 1500 c.c. as some insist, these great periods since these "ape-men" existed must be enormously increased, in some cases 50%.

14. If, on the other hand, the pithecanthropus really lived 750,000 years ago, what, with normal development, should have been its skull capacity, if life began 60,000,000 ago? Ans. 98.75%; or 1481 c.c. If life began 500,000,000 years ago? Ans. 99.85%; or 1497.77 c.c. In either case, practically normal.

15. If the Piltdown "man" lived 150,000 years ago, as claimed, what should have been his brain capacity, if life has lasted 60,000,000 years? Ans. 99.75%; or 1496.25 c.c. If 500,000,000 years? Ans. 99.97%; or 1499.55 c.c. Very nearly normal.

The above problems prove either that these alleged links could not have lived in the periods assigned them, or else they must have had a brain capacity almost normal, and far greater than assigned to them.

16. The habitable countries of the world-total 50,670,837 sq. mi. If we estimate that the garden of Eden occupied 10,000 sq. mi. or 6,400,000 acres, there would be 5067 such areas in the world. What chance would Moses have, not knowing, to guess the correct location? Ans. 1 chance out of 5067,—virtually none at all.

17. If Moses, not knowing the order of creation, enumerates 11 great events in their correct scientific order, what chance had he to guess the correct order? Ans. I chance out of 39,916,800. If 15 great events, as some biblical scholars point out? Ans. I chance out of 1,307,674,368,000. (Solve by Permutation.)

18. If there are now 1,500,000 species of animals, coming from a single primordial germ or cell which existed 60,000,000 years ago, how many species of animals should have arisen or matured in the last 6000 years? Ans. 3000; or one every two years. If life has existed 500,000,000 years, 360 new animal species were due in the last 6000 years. Evolutionists declare they do not know that a single new species has arisen in the last 6000 years! Even Darwin said, "Not one change of species into another is on record."

19. If the skeletons of 200,000 prehistoric horses were found in a single locality, Lyons, France, how many skeletons of prehistoric man should we expect? Ans. Many millions. How many are there? Not a single or undisputed skeleton of an ape-man!

20. If each of the two eyes and ears as well as the nose and the mouth occupy, on an average, one-thousandth part of the surface of the body, what, if we exclude God's design, is the mathematical probability that they would appear where they are? Ans. .OO1 x .OO1 x .001 x .001 x .001 x .001; =.000,000,000,000,000,001; or 1 chance in a billion billion! (Solved by Compound Probability.)

21. Evolutionists claim at least 8 great transmutations from matter to man: matter, plant-life, invertebrates, vertebrates, fishes, amphibians, reptiles, mammals and man. If we make the extremely generous estimate of 60% to represent the probability of each transmutation, what is the compound probability that all would take place? Ans. 1 chance in 60, which means an extreme improbability.

22. If there is 1 chance in 10 that each transmutation has taken place, which is far more than the evidence warrants, what fraction represents the probability that all these great changes have occurred? Ans. .1 raised to the eighth power, or .00000001; or 1 chance in 100,000,000.

23. If the probability of a change of one member of one species into another species be expressed by .1 (an over-estimate), what fraction marks

the probability of a million members making the same change? Ans. .1 raised to the millionth power; or 1 preceded by 999,999 decimal ciphers; or a common fraction with 1 as a numerator and a million figures as a denominator; or 1 chance out of a number expressed by 1,000,000 figures, which would fill 3 volumes like this book. Such changes were absolutely impossible, but necessary for evolution.

24. If the scattered remains of the pithecanthropus were found in the sand only 40 ft. below the surface, and the rate of accumulation were no greater than the slow accretions that buried the mountain city of Jerusalem 20 feet deep in 1900 years, what would be the extreme age of these remains? Ans. 3800 years, instead of 750,000 years.

25. If the Heidelberg jaw was found in sand 69 ft. deep, what would be its maximum age, estimated in the same way? Ans. 6555 years instead of 375,000. Who believes that sand in a river valley would accumulate no more rapidly than dust on the mountains? Or that it took 750,000 or even 375,000 years to cover with sand these precious remains such a shallow depth? A few centuries at most would account for such a depth. Can there be any doubt that these were abnormal bones of historic man and brute?

26. Did any other false theory that ever posed as science, have less to support its claims than evolution?

27. Believing that a Christian should give to the Lord all above his necessities, none of the profits on this book will be retained by the publisher, but all will be donated to missions, to be perpetually loaned to churches, and to preach the gospel through the secular newspapers, of the world, and to aid in the free distribution of this book as explained on pages 116 and 117. How many churches will every $1000 together with the compound interest thereon, help to build in 300 years, if the average loan to each church is $1000 for 5 years at 5%? Ans. 8,229,024; and the new principal will then be $2,273,528,000.

28. How could $1000 be given to do more good than for these three purposes?

29. "For what shall it profit a man, if he shall gain the whole world and lose his own soul?"

30. What shall it profit a man, if he wins great fame as a scientist, persuades a great multitude to accept evolution, infidelity and atheism, and leads a great company to the lost world, by destroying their faith in God and in Jesus Christ?

50
THE SUPREMACY OF JESUS

From far-off Australia comes this sermon by Rev. R. Ditterich. What more fitting climax in honor of Christ, whose worshipers belt the globe? "Christ is All," a pean of praise, which has been sung both sides the sea, and published in three Hymnals and over sixty song books, will close this volume, dedicated to the glory of God.

Text: "Thou art the Christ, the Son of the living God."—Matt. 16.16.

Jesus asked a great question, and Peter made a great reply. No prophet, no priest, no king, no patriarch of Israel had ever been greeted in such fashion. Of nobody else in the world are these words spoken today. How pure must have been the life, how majestic the personality, how wise the utterances, how divine the deeds, that compelled this thrilling answer from the apostle's lips. Surely something really wonderful beyond all previous Hebrew experience was necessary before Jews could bring themselves to acknowledge any man, however exalted, as divine. The miracle of winning such a confession is testimony to the sovereign greatness of Jesus.

We, too, have to answer the same question, and there are facts which lead us to the same great confession of faith.

FIVE TREMENDOUS FACTS

1. Jesus, a peasant, is hailed today as King by people speaking 750 languages and dialects, in all climes, and of all classes. People of every color raise to Him the song of praise and crown Him "Lord of all." There is nothing like this in all history. No other has ever approached this degree of sovereignty. His kingdom pervades the world. It is a fact that challenges thought. No world conqueror has ever had such an empire. Beside this the royalty of men like Alexander, Caesar, Charlemagne, Napoleon, and more modern aspirants is shadowy and ghostlike. His is an abiding and a spiritual dominion.

2. Though an unlettered peasant, Jesus has become the world's greatest teacher. For all our best knowledge of God, for the revelation of divine Fatherly love, for our highest ideals of virtue, for man's most glorious hope, people on all sides look to Him. Not only men of the highest rank, but men of the richest culture sit at His feet. The purest souls sit at His feet. His golden rule will never be supplanted. His name has become the synonym for all that is true and gracious. To be Christ-like must ever remain man's highest ideal.

3. He was a Jew, and yet He founded the brotherhood of man. In His day Jews had no dealings with Samaritans. But Jesus had. Jews were fenced off from all other nations in the most exclusive way. But His heart was all-inclusive, and He broke down all walls that separated class from class as well as nation from nation. His thought was universal. His spirit was international. He founded a kingdom based, as Napoleon said, not on force but on love, and love is universal. It leaps over mountains, it spans oceans. It speaks in all tongues. The true League of Nations and the real disarmament are part of His plan for the world. He was son of Israel only incidentally. Essentially He was Son of Man—the true brother of all mankind.

4. His life was short, but it changed the world. No one ever did so much in so short a time. At the most his years numbered thirty-three years, and of these only a little less than three were devoted to public ministry, and these were spent in a conquered province of the Roman Empire. He was killed by aliens at the request of His own countrymen. And yet time is reckoned from His birth. The very terms B.C. and A.D. have great significance. He divides not only time, but also space. The nations are Christian and non-Christian, which is about equal to saying, civilized and barbarous. One has only to think of the ideals and practices of pagan people before they received the influences of Christianity to see the difference He makes everywhere. No tribe on earth was ever lifted from savagery by the influence of Socrates, no crime-soaked soul was ever saved by his name and yet Socrates was the wisest and noblest of the Greeks. He lived for seventy years and for forty years taught the young men in the most cultured age and among the most intellectual people in the world. But Jesus has lifted cannibals and washed the souls of men who were steeped in blackest vice. The rationalist Lecky said that the simple record of His three brief years of active life had done more to regenerate and soften mankind than all the disquisitions of philosophers and than all the exhortations of moralists.

5. He was crucified, and made of the cross a throne from which to rule the hearts of men. The cross was a gallows far more hideous and cruel than the hangman's gallows. It was the symbol of crime, of shame, of degradation. He transformed it. It is today the symbol of love, of purity, of virtue. His dream came true. Once only did a man dream that by dying upon a cross would He teach men to say that God is love, that love is universal, that there is hope for sinners, and that the worship of God must be spiritual. This is the miracle of the ages. The Crucified has become the King.

Here then are five tremendous facts. They are unique. If only one were true it would make Him remarkable, but they are all true.

THE MEANING OF THE FACTS

What shall we say of this Man? He accepted Peter's tribute. He allowed Jews to take up stones to stone Him for claiming to be Son of God. He was conscious of being divine. He forgave sins, which is God's prerogative. He promised rest to the weary soul, which the Old Testament set forth as God's own gift. He said that He came to give life eternal, although God is the giver of life. He said that none could know the Father except through Him. He spoke to God of the glory which they shared together before the world was. Just in proportion as men have acknowledged His claims in their hearts have they found peace with God and conquest over sin and the fear of worldly evil. As we consider all these things we are led to repeat Peter's confession, "Thou are the Christ, the Son of the living God," for God the Father's face shines upon us through Him and heaven is opened to us as we look upon Him. In the heart of this the purest of men was the clear, constant consciousness that He was divine. He always spoke and acted consistently with this consciousness. Unique in character, He made claims that would have stamped any other man as an impostor. Humility and majesty dwell together in Him. He could say, "I am meek and lowly in heart," and also "I and my Father are one." He would call men His "brethren" and yet accept from them the words, "My Lord and my God." This wonderful character came of a race that had for ages looked for the coming of a Messiah, and whose prophetic literature was burdened with this hope. After his death his disciples who were heartbroken and cowed became inspired with a heroism that cheerfully faced martyrdom. All these facts are shining lights that point to the truth which Peter confessed. That truth is enshrined in the triumphant words of the Te Deum, "Thou are the King of glory, O Christ. Thou art the everlasting Son of the Father."

And the Christ of history, the exalted Son of God, is a living Presence with us today. Not remote but ever near, He walks by our side in all life's experiences. Not only enthroned in heavenly glory

> "But warm, sweet, tender, even yet
>   A present help is He,
> And faith has still its Olivet
>   And love its Galilee."

Such is our wonderful Saviour, a Friend with human heart of sympathy who has trod our pathway and is touched with the feeling of our infirmities; a Shepherd who gave His life for the sheep in an all-atoning sacrifice; an Advocate who represents us with all-prevailing power before the throne of the Judge Eternal; a Champion who Can break the power of canceled sin and set the prisoner free; a Victor who can smite death's threatening wave before us; a Lord in whom we see the beauty and glory of the face of God. We are called upon to confess Him with lip and life. To us to live is Christ. Knowing Him we have eternal life. We have all the soul needs in Jesus. There is no substitute for Him. None can share His throne in our hearts. The Kingdom is His who is the Christ—the anointed King. Our joy is in Him, where all fullness dwells. We can say with Charles Wesley, "Thou, O Christ, art all I want," and our daily life should be one of close, constant communion with Christ.

Christ Is All.

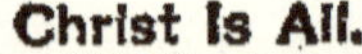

W. A. WILLIAMS, by per.

No. 21. CHRIST IS ALL.

"Unto you therefore which believe he is precious." —Pet. 11:7.

W. A. WILLIAMS, by per.

1. I entered once a home of care,
For age and penury were there,
Yet peace and joy withal;
I asked the lonely mother whence
Her helpless widowhood's defence.
She told me, "Christ was all."
Christ is all, all in all,
She told me "Christ was all".

2. I stood beside a dying bed,
Where lay a child with aching head,
Waiting for Jesus' call,
I marked his smile, 'twas sweet as May,
And as his spirit passed away,
He whispered, "Christ is all."
Christ is all, all in all,
He whispered "Christ is all."

3. I saw the martyr at the at the stake,
The flames could not his courage shake,
Nor death his soul appall,
I asked him whence his strength was giv'n,
He looked triumphantly to Heav'n,
And answered "Christ is all."
Christ is all, all in all,
He answered, "Christ is all."

4. I saw the gospel herald go,
To Afric's sand and Greenland's snow,
To save from Satan's thrall:
No home nor life he counted dear,
Midst wants and perils owned no fear.
He felt that "Christ is all."
Christ is all, all in all,
He felt that "Christ is all."

5. I dreamed that hoary time had fled;
The earth and sea gave up their dead,
A fire dissolved this ball;
I saw the church's ransom'd throng,

I heard the burden of their song.
'twas "Christ is all in all."
Christ is all, all in all,
'Twas Christ is all in all.

6. Then come to Christ, oh! come today.
The Father, Son, and Spirit say;
The Bride repeats the call;
For he will cleanse your guilty stains,
His love will sooth your weary pains,
For "Christ is all in all."
Christ is all, all in all,
For "Christ is all in all."